1 MONTH OF FREE READING

at

www.ForgottenBooks.com

By purchasing this book you are eligible for one month membership to ForgottenBooks.com, giving you unlimited access to our entire collection of over 1,000,000 titles via our web site and mobile apps.

To claim your free month visit:

www.forgottenbooks.com/free6095

* Offer is valid for 45 days from date of purchase. Terms and conditions apply.

ISBN 978-0-365-20207-3
PIBN 10006095

This book is a reproduction of an important historical work. Forgotten Books uses state-of-the-art technology to digitally reconstruct the work, preserving the original format whilst repairing imperfections present in the aged copy. In rare cases, an imperfection in the original, such as a blemish or missing page, may be replicated in our edition. We do, however, repair the vast majority of imperfections successfully; any imperfections that remain are intentionally left to preserve the state of such historical works.

Forgotten Books is a registered trademark of FB &c Ltd.
Copyright © 2018 FB &c Ltd.
FB &c Ltd, Dalton House, 60 Windsor Avenue, London, SW19 2RR.
Company number 08720141. Registered in England and Wales.

For support please visit www.forgottenbooks.com

THE ELEMENTS
OF
ANIMAL BIOLOGY

HOLMES

THE ELEMENTS

OF

ANIMAL BIOLOGY

BY

S. J. HOLMES, PH. D.

PROFESSOR OF ZOOLOGY, UNIVERSITY OF CALIFORNIA

WITH 249 ILLUSTRATIONS

PHILADELPHIA
P. BLAKISTON'S SON & CO.
1012 WALNUT STREET

COPYRIGHT, 1919, BY P. BLAKISTON'S SON & CO.

PREFACE

This volume is intended as an introduction to the elements of animal biology for the use of students in the high school. The cut-and-dried method of exposition which is so commonly found in text-books and which so frequently deprives them of all traces of stimulating quality has been avoided so far as was deemed compatible with the presentation of such subject matter as a text-book should contain. Although the book would best fulfil its purpose if read in connection with laboratory work, I have not included directions for such work, partly because it would add considerably to the bulk of the volume, but chiefly because so many teachers nowadays prefer to make laboratory outlines of their own.

The order in which the main topics are treated is essentially like that which is followed in several of the best recent text-books. The general experience of teachers of biology has shown it to be eminently desirable that the student should possess a general knowledge of the animal kingdom as a preparation for the study of physiology. The section on the elements of physiology has therefore been placed after the part devoted to a survey of the principal groups of animals. Rather more than the usual amount of attention is given to the rôle of bacteria in causing disease, and to the way in which diseases are spread and how they may be avoided.

The third part of the book dealing with general topics such as evolution, heredity and eugenics begins with a brief survey of the phenomena of sex and reproduction in the chapter on the Perpetuation of Life. No reference is made in the text to the much discussed problems of sex hygiene. It was thought desirable to leave this topic to the discretion of the teacher to be handled in whatever way he or she considers to be most effective. An attempt, however, has been made to supply the students with a general basis of knowledge of the facts of reproduction and development so that a little added instruction of the right sort will aid them in avoiding the dangers and pitfalls into which ignorance is continually leading so many of our youth. Sex hygiene should be associated not merely with considerations of personal welfare, but with the welfare of future generations. It is highly important that the youth of to-day who are to be the parents of to-morrow should be imbued with a sense of their obligations as fathers and mothers of children. They need to be made aware that it is a matter of great moment what kind of people are supplying the larger part of our future population. And an effort has accordingly been made in the discussion of heredity and eugenics to prepare them for an appreciation of the importance of a knowledge of the forces that are working toward the improvement or the deterioration of the inborn qualities of the race.

In the preparation of this book I have profited by the criticism of several of my colleagues. I am indebted to Dr. W. W. Cort for reading the entire manuscript, to Dr. J. Grinnell for reading the chapters on birds and mammals, to Mr. Tracy Storer for reading the chapters on reptiles and amphibians, to Dr. J. F. Daniel for reading the chapter on fishes, and to Dr. E. C. Van Dyke for reading the chapters on insects. Dr. L. J. Cole of the University of Wisconsin has had the exceptional patience to read several of the chapters in the almost illegible script in which they were originally written. I wish to thank Prof. J. S. Kingsley and Henry Holt and Co. for permission to use figures from Kingsley's translations of Hertwig's Zoology. Dr. Paul Carus has very kindly permitted me to use several cuts from Romanes' *Darwin and After Darwin* issued by the Open Court Publishing Company. To Dr. B. W. Evermann I am indebted for photographs of several groups of mammals and birds in the museum of the California Academy of Sciences. Most of all I am indebted to my wife for help in many ways during the preparation of this work.

S. J. H.

BERKELEY, CALIF.

CONTENTS

PART I.—THE ANIMAL KINGDOM.

CHAPTER.	PAGE
I. The Grasshopper and other Orthoptera	1
II. How Animals are Classified	17
III. The Lepidoptera or Butterflies and Moths	20
IV. The Hemiptera, or the Bugs and their Allies	31
V. The Diptera or Flies	40
VI. The Coleoptera or Beetles	51
VII. The Hymenoptera or Bees, Ants, Wasps and their Allies	56
VIII. The Dragon Flies, Damsel Flies, May Flies, Stone Flies and Caddis Flies	70
IX. The Myriapods and Arachnids	75
X. The Crayfish and Other Crustacea	84
XI. The Mollusca	98
XII. The Echinoderms	108
XIII. The Ringed Worms or Annelids	114
XIV. The Round Worms and Flat Worms	122
XV. The Coelenterates and Sponges	130
XVI. The Protozoa or the Simplest Animals	141
XVII. The Lowest Vertebrates and their Nearest Allies	153
XVIII. The Fishes	158
XIX. The Amphibia (Frogs, Toads, Newts and Salamanders)	169
XX. The Reptiles	176
XXI. The Birds	183
XXII. The Mammals	209

PART II.—THE ELEMENTS OF PHYSIOLOGY.

XXIII. The Chemical Basis of Life	232
XXIV. Cells and Tissues	240
XXV. Digestion	245
XXVI. Foods and Their Uses	252

Chapter.	Page
XXVII. The Blood and Circulation	259
XXVIII. Respiration	269
XXIX. Excretion	277
XXX. Internal Secretions and the Ductless Glands	280
XXXI. The Skin	283
XXXII. The Skeleton and the Muscles	287
XXXIII. The Nervous System	294
XXXIV. The Organs of Sense	304
XXXV. Alcohol and Tobacco	311
XXXVI. Bacteria and Disease	317

PART III.—GENERAL FEATURES AND ADAPTATIONS.

XXXVII. The Perpetuation of Life	331
XXXVIII. The Evolution of Life	341
XXXIX. Divergence and Adaptation	362
XL. Heredity and Human Improvement	370

PART I
THE ANIMAL KINGDOM

ANIMAL BIOLOGY

CHAPTER I

THE GRASSHOPPER AND OTHER ORTHOPTERA

Among the most common things with which we come into contact in this world are living beings, and it is therefore highly desirable that we know a good deal about them. The science which deals with the living world is called Biology, and it falls into two subdivisions, one, Botany, which is concerned with plants, and the other, Zoology, which treats of animals. No one, be he ever so dull, can escape picking up some information on the subject matter of these branches of science, yet comparatively few realize the great importance and interest which attaches to the study of the world of life. Most of us are blind to countless wonderful and beautiful things which a little well-directed observation would disclose to our view. And we often shrink from many forms with a feeling of repugnance which a closer acquaintance would change to one of interest and admiration for the remarkable adaptations of their structure and activities.

To begin our study of animal life, we shall select a common and familiar form, the grasshopper. All living creatures have much in common, and what we learn about one kind will help us very materially in learning about others. The grasshopper is a member of the great class of insects which is the largest group in the animal kingdom. It is a creature of complex structure, but formed upon a pattern very different from that of our own body. Like ourselves, however, it can walk, jump, eat, breathe, see, hear, smell, make sounds and perform a great number of other activities which show that it is not so different from a human being as we might at first be disposed to believe.

In order to understand how these activities are carried on it is necessary to study briefly the grasshopper's structure. The skeleton of the grasshopper's body is on the outside, instead of on the inside as in ourselves, and it is composed of a substance called *chitin*, which is thickened in some regions where it has acquired a certain rigidity. Externally the body shows a division into rings marking

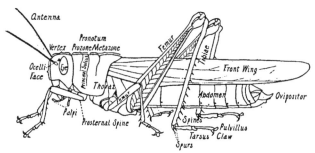

FIG. 1.—Side view of a typical grasshopper. (After Woodworth.)

the individual *segments* of which the body is built up. Some of the segments can move one upon the other. In these cases the chitin between the segments is thinner than elsewhere so as to become flexible, thus allowing freedom of movement. In other cases the segments are so closely united that there is no motion between them.

The body of the grasshopper shows a division into three parts, the *head, thorax* and *abdomen*. Look with a hand lens at the large eyes at the sides of the head and you may see that they present a finely checkered appearance due to their being composed of smaller elements. The grasshopper's eye is compound, and when we look at the transparent outer covering, or *cornea*, of the eyes with a microscope it will be seen to be divided into a number of six-sided areas. Each of these areas lies over a sort of simple eye; in fact the compound eye may be regarded as composed of a large number of simple eyes lying side by side, the whole forming a very efficient organ of sight as you may easily convince yourself by trying to catch grasshoppers in the field. Besides the compound eyes there are three simple ones or *ocelli* at the top of the head, but little is known concerning their precise use.

The long feelers or *antennæ* on the front of the head are composed of many movable segments. By watching a live grasshopper you may discover indications that the antennæ are used as organs of touch. They are also, curiously enough, organs of the sense of smell. The microscope shows that they contain numerous pits, the *olfáctory pits*, which are probably organs for detecting odorous substances. On the mouth parts of the grasshopper there are certain organs resembling short feelers, called the *palps*, which the grasshopper uses considerably when feeding. These palps have been thought to contain organs of taste, but it is probable that taste organs occur further within the mouth also.

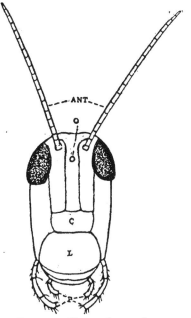

FIG. 2.—Face of grasshopper. *ANT*, antennæ; *C*, clypeus; *L*, upper lip; *O*, ocellus; *P*, palpi.

The grasshopper is furnished with three pairs of mouth parts: (1) a pair of strong jaws or *mandibles;* (2) a pair of smaller accessory jaws, the *first maxillæ* and (3) the *labium* or *lower lip,* which is composed of the *second maxillæ* more or less fused together. The peculiar working of these organs may be readily seen in a living specimen.

The *thorax,* or the part of the body immediately behind the head, consists of three segments called the *prothorax,* *mesothorax* and *metathorax* in order from before backward. Like nearly all insects the grasshopper has six legs, a single pair being attached to each segment of the thorax. The first two pairs which are fitted for walking or climbing are quite different from the last pair which is mainly used for jumping. It may readily be seen that the parts of the legs are united by flexible membranes at the joints so as to permit of free movement. Each leg consists of two short segments near the base, a relatively large part, the *femur,* followed by the *tibia,* and finally the foot, or *tarsus.* The latter ends in a pair of claws and is furnished wih pads below the segments. The hind legs are especially noticeable on account of their large size and the double row of sharp spines on the posterior side of the tibia. The function of these spines may readily be discovered by observing live grasshoppers.

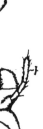

FIG. 3.—Mouth parts of grasshopper. *m,* mandible; *max,* maxilla; *l,* labium; *p,* palpus.

The first pair of wings which are situated on the mesothorax are often called the wing covers as they lie over the large second pair of wings which are the chief organs of flight. The first wings are narrow, relatively thick and usually of a dull color. The second pair are broad, thin of texture and frequently brightly colored. Ordinarily the latter are not seen except during flight as they lie folded up under the first pair. Both pairs of wings may be regarded as folds of the general chitinous covering of the body. Consequently they consist of a double chitinous membrane. In certain places the chitin is thickened to form the so-called *veins* which serve to give the wings the degree of rigidity necessary for flight.

The abdomen is composed of segments most of which are similar in structure and freely movable upon one another. The upper and lower parts of the segments are united by a flexible membrane so as to permit a certain amount of vertical movement between these parts. At the posterior end of the abdomen is the opening of the intestine and certain appendages used in reproduction. The end of the abdomen differs in the two sexes and affords an easy means of distinguishing the male from the female. In the female the tip of the abdomen is furnished with two pairs of acute processes forming an organ called the *ovipositor* which is used in laying the eggs in the ground. In the male the end of the abdomen is blunt and swollen and is entirely devoid of an ovipositor.

FIG. 4.—Part of a tracheal tube with coating of cells.

One of the most peculiar features of the grasshopper's life is its mode of breathing. Watch the extension and contraction of the abdomen and the changes in shape of the segments. We might easily conjecture that these movements had to do with breathing, but it is not so apparent where air is taken into and expelled from the body. With a hand lens, however, one may see a number of small apertures called the *spiracles* along the sides of the abdomen, and two larger pairs on the thorax. These lead to the breathing tubes or *tracheæ* which ramify throughout the body and carry air to all the internal organs.

The grasshopper also differs greatly from most animals

in the location of its organ of hearing. Look on the side of the first segment of the abdomen and you will find an oval aperture covered with membrane; this is furnished with a delicate apparatus which is stimulated by the vibrations of the air and thus makes its possessor aware of the presence of sound. As animals which hear usually have some means of making a noise themselves, we find that grasshoppers are frequently furnished with an instrument for the production of song, although it is present only in the male sex. This instrument consists in most cases of a series of fine teeth on the inner side of the femur. As this apparatus is rubbed over one of the veins of the wing cases it produces a shrill note. If one carefully approaches a singing grasshopper he may see the femur drawn across the wing cover, producing a sound much as the violinist does by drawing his bow over a string. The musical sound probably serves, like the songs of birds, as a means of bringing the sexes together.

After this study of the external parts of the grasshopper let us consider briefly some of the features of its internal structure. The part of the body concerned with the digestion of food is the *alimentary canal* which is a tube leading from the mouth to the posterior end of the body. Different parts of this tube differ in structure and in function. Leading from the mouth is a short and narrow division called the *esophagus*. Connected with a prominence at the anterior end of the esophagus are the *salivary glands* whose function is to secrete and pour into the alimentary canal a fluid which aids in digestion. Posteriorly the esophagus leads to a *crop* lined internally with rows of chitinous teeth which probably serve to grind up the food. Behind the crop is a large, thin-walled *stomach* at the anterior end of which open a number of tubes called *gastric cœca*. The stomach passes into the *intestine* which is narrow and more or less coiled. The latter leads to the terminal portion of the alimentary canal, the *rectum,* which opens through the last segment of the body. The food as it passes down the alimentary canal is acted on by the saliva, the secretion of the gastric cœca, and other fluids formed by the walls of the stomach whereby it undergoes a process of digestion after which the soluble materials are absorbed through the walls of the alimentary canal, especially the stomach and intestine, and carried to various parts of the body. The undigested residue is expelled through the rectum.

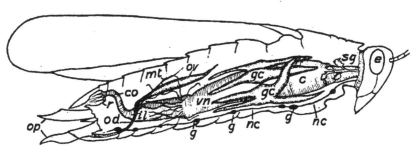

FIG. 5.—Diagram of the internal organs of a grasshopper. *c,* crop; *gc,* gastric cœca; *vn,* ventriculus, *il,* ileum or anterior part of the intestine; *co,* colon, *r,* rectum; *e,* eye; *g, g,* ganglia; *nc,* nerve cord; *sg,* salivary glands; *mt,* Malpighian tubules; *ov,* ovary; *od,* oviduct; *op,* ovipositor. (Modified from Brooks.)

The absorbed food materials are carried to different organs by means of the blood which is not red as in ourselves, but nearly colorless. The organ for propelling the blood is the heart, an organ very different in appearance and position from our own heart, as it consists of a long tube lying along the upper part of the abdomen. It is closed behind but open in front, and is perforated by several pairs of openings along the sides. Blood enters through these lateral openings and is prevented from flowing back by valves. The heart beats or contracts from behind forward so that the blood which is drawn in through the sides is expelled at the front of the heart through short blood vessels. There are almost no well-defined blood vessels in the grasshopper, and the blood, after being forced out of the short vessels at the front of the heart, passes into irregular spaces between the tissues and organs of the body until it finds its way back to the heart again, when it is taken in through the lateral openings and sent out on a new journey. In many insects, especially the transparent young or larval stages of aquatic species, it is possible to see the beating of the heart and the flow of blood in the living organism.

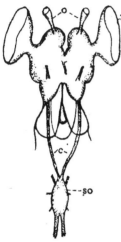

FIG. 6.—Brain of grasshopper from in front. C, commissures around the esophagus; E, nervous supply of eye; O, nerves to ocelli; SO, subesophageal ganglion.

The organs in the grasshopper which correspond in function to the human kidneys and which therefore serve as a means of getting rid of certain waste products in the blood consist of a series of fine thread-like tubules, called *Malpighian* or *urinary tubules*, which empty into the intestine close to where it joins the stomach. The waste matter collected by these tubules therefore passes out of the body with the undigested portions of the food. Other waste products, especially carbon dioxide, are removed by the tracheal tubes.

The brain of the grasshopper is situated in the upper part of the head above the esophagus. From the brain nerves go to the eyes, ocelli and antennæ, and from the lower side a nerve cord passes on either side of the esophagus to a nerve mass called the *subesophageal ganglion* which supplies nerves to the mouth parts. This ganglion is the first of a series of paired ganglia extending along the ventral side of the body. These ganglia are connected by a double nerve cord, the two parts of which lie very close together and appear in the abdomen as a single strand. There is a paired ganglion in each of the three segments of the thorax and five ganglia in the abdomen. From these ganglia nerves are given off to the muscles and sense organs of neighboring parts.

Movements of parts of the body are effected by means of muscles. The thorax contains muscles of unusually large size which are used for moving the wings. Muscles act by contracting and thereby producing movement in the parts upon which they are inserted. The impulses causing the contraction are conveyed by nerves which pass to the muscles from the ganglia.

The *ovaries*, or organs for producing the eggs, are situated in the abdomen of the female. They present the appearance of paired masses of eggs in various stages of growth from an exceedingly minute size to the full grown egg. When the eggs have attained their full size they are discharged from the ovary into a tube, the *oviduct*. A short distance from the posterior end of the body the two oviducts unite to form a median duct which opens between the bases of the valves of the ovipositors previously described. During their passage down the oviduct the eggs become surrounded by a sticky substance secreted by certain glands which open near the junction of the two oviducts. It is this sticky substance which causes the eggs to adhere in masses after they are laid and which subsequently protects them from the injurious effects of moisture.

The male organs corresponding to the ovaries are called the *spermaries* or *testes*. They lie in the abdomen above the intestine and are so closely united that they appear as a single organ. They give rise to slender ducts which open at the posterior end of the body. The spermaries produce minute bodies called *spermatozoa* whose function it is to unite with or fertilize the eggs and thereby render them capable of development.

The eggs are fertilized before they are laid. In the late summer or fall the female bores with her abdomen a shallow hole in the earth and deposits her mass of eggs which lie there over winter and hatch out in the following summer. The young grasshopper at its first appearance

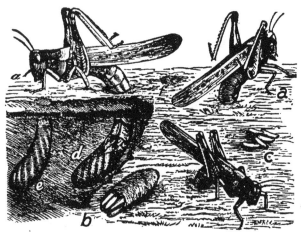

FIG. 7.—Grasshoppers laying eggs. *a, a, a*, female in different positions. *b*, egg pod, *c*, separate eggs, *d, e*, earth removed to expose the pods. (After Riley.).

upon the stage of life is conspicuously different from the adult in several respects; it is small in size, soft bodied, entirely devoid of wings, and provided with a head which seems all out of proportion to its diminutive body. It starts at once on the main business of its early life which is eating, eagerly devouring all sorts of plant life and consequently growing rapidly. As a result of its growth the chitinous skin or exoskeleton which is made of comparatively inelastic and unyielding material becomes too small. Then comes the process of *molting* or shedding the skin.

The skin splits down the middle of the back, the abdomen is drawn forward out of its old case, the legs are pulled out of their coverings and the grasshopper slowly pulls itself out of all of its old clothes and appears in a new but thin external covering which had been forming preparatory to casting off the older one. The recently molted grass-

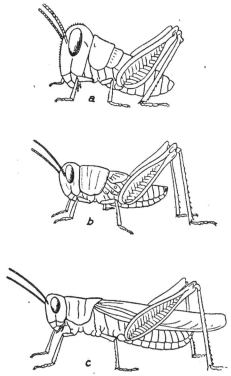

Fig. 8.—Three stages in the metamorphosis of a grasshopper. (After Herms.)

hopper appears quite suddenly larger than before. The new skin hardens, the grasshopper eats and grows, and before long it has to undergo a new molt. The skin is shed four or five times before the grasshopper reaches maturity. During this series of molts gradual changes of form, which are spoken of as *metamorphosis*, take place. One of the most salient features of this process is the development of the wings which first appear as small folds on the posterior margins of the second and third segments of the thorax. The wings become larger with successive molts and acquire a joint, or articulation, at the base which renders them freely moveable. Forms such as the grasshopper and its allies in which the newly hatched young resembles the adult and passes into the latter by gradual stages are said to undergo an *incomplete metamorphosis*. Insects with a *complete metamorphosis* pass through stages marked by abrupt and extensive changes. In these forms the larva passes into a usually quiescent *pupa* from which finally emerges the *imago* or adult insect.

The description of a grasshopper previously given will apply to a large number of the more typical kinds. There are several hundred different species of grasshoppers, and they occur in almost all countries of the tropical and temperate regions of the earth. We commonly find considerable differences between the grasshoppers of different countries, and any one country usually contains several species. Grasshoppers are very common in meadows and grain fields. The Carolina locust, or roadside grasshopper, frequents roadsides and other bare patches of ground where it is difficult to detect on account of the similarity of its color to that of the soil. Many other species are protectively colored when in their natural surroundings. Certain species of grasshoppers are migratory and some of the most destructive forms belong to this group. In some of the migratory species the air sacs connected with the tracheæ are well developed and when inflated with air serve to buoy the insects up during long flights. These forms fly in swarms which sometimes consist of incredible numbers. In 1889 a swarm which passed over the Red Sea was over 2000 square miles in area. The swarms are sometimes so dense that the sun is darkened and when they alight they eat up within a short time almost every green plant within their reach. Plagues of locusts in the old world have been frequently recorded, some accounts of them being found in the Bible.

The most destructive of the grasshoppers of our country is the celebrated Rocky Mountain locust which during a series of years 1874-6 caused enormous damage in Kansas, Nebraska, Missouri and other western states. Old inhabitants will never forget the grasshopper years. The insects came down upon them in such vast swarms

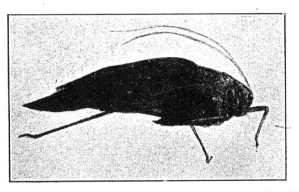

Fig. 9.—The California katydid, *Scudderia furcifera* Scudd. Male, natural size. (After Essig.)

that in many places they ate up completely the crops of corn and grain, destroyed the pastures, stripped many shrubs of their foliage and devoured the weeds, even down to such unsavory ones as smart weed and dog-fennel. Stock thus deprived of their food perished in large numbers and the inhabitants underwent great hardships on account of their losses. Professor Packard estimates that the losses due to grasshoppers during four years amounted to $200,000,000. Very few of our common grasshoppers are so destructive as the Rocky Mountain locust although several species do considerable damage. One of the most common and widespread pests is the red-legged grasshopper which is closely related to the preceding species.

Grasshoppers have been held in check by various methods. Plowing the ground destroys large numbers of the eggs as they seldom develop if covered with a few inches of soil. Grasshoppers are sometimes poisoned by bran mixed with Paris green or London purple. A good deal of success in destroying them has been attained by dragging over the ground machines called "hopperdozers" in which the grasshoppers are collected and killed by kerosene oil. This substance is a deadly poison to grasshoppers and most other insects, a slight contact with it usually proving fatal.

FIG. 10.—Eggs of the angular-winged katydid attached to twigs. The holes have been made by the egg parasite, *Eupelmus mirabilis* (Walsh). (After Essig.)

Closely allied to grasshoppers such as we have described are the so-called long-horned grasshoppers, katydids and their allies, all of which have long, slender antennæ. The fore wings in the males are furnished with a peculiar apparatus at the base, by which they can make a noise when the wings are rubbed together. Only the males of the katydids and grasshoppers sing. The females have an ovipositor, frequently very long, by which eggs may be affixed to or inserted into the stems of plants.

Other relatives of the grasshoppers are the crickets with whose dark and glossy bodies, lively movements and cheerful chirping we are all more or less familiar. In these forms also it is the male that makes the song and he does it by rubbing together the sound-producing organs on the bases of the fore wings. The female may be distinguished by the absence of these organs and by the presence of an ovipositor which frequently is of considerable length. The white crickets which are usually found upon plants, and the mole cricket which lives under ground and has its fore legs modified for burrowing are less typical members of the cricket family.

Less popular relatives are the cockroaches, some species of which are found under stones and logs while others prefer to live within houses where they devour all sorts of

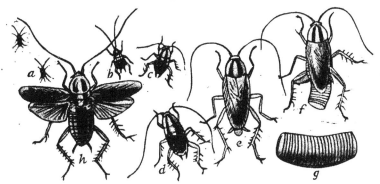

FIG. 11.—A common cockroach, *Blattella germanica. a-e*, various stages of development, *f*, female carrying egg case; g, egg case. (After Riley.)

food-stuffs and make themselves a general nuisance. They are very lively rascals, carrying on most of their depredations at night, although they are often seen during the day. Their flattened bodies enable them to crawl into narrow crevices which afford them concealment. The eggs are laid in a very peculiar egg case which is carried around for a while by the female.

More distant kin of the grasshoppers are the walking-sticks and mantids. The former are remarkable for their long, narrow body and slender legs, the whole insect being readily mistaken for a twig. The leaf insect of the tropics is allied to the walking-stick, but it is furnished with green wings whose veins closely resemble those of a leaf. The mantids have a long prothorax and strong fore legs

FIG. 12.—Walking-stick.

fitted for seizing other insects on which they feed. The fore legs are held in a devout attitude which doubtless suggested the name "praying mantis" and gave

FIG. 13.—American mantis, *Mantis carolina.*

rise to the many superstitions associated with this insect. The so-called attitude of prayer is really one of readiness for quickly grabbing any insect within reach.

CHAPTER II

HOW ANIMALS ARE CLASSIFIED

The various insects we have described resemble one another in different degrees so that it is possible to arrange them in groups according to their degrees of likeness. This process of grouping is called *classification*. For many purposes it is desirable to divide various kinds of objects such as books into classes and this is especially desirable in the case of animals and plants of which there are such vast numbers of different kinds. Animals or plants which very closely resemble one another, and which are commonly spoken of as belonging to the same kind constitute what is called a *species*. For instance, the individuals of the Rocky Mountain locust form one species; those of the red-legged locust another. The Grizzly bear is one species, the black bear another and the European brown bear a third. While species is the last or smallest group usually dealt with, we sometimes recognize smaller groups within the same species, which are called *varieties*. We commonly speak of varieties of corn, wheat, cattle, pigs, etc., where the groups differ but slightly, and where one is known to have been derived from the other, or where one group shades into the other one. It is not possible to draw a sharp distinction between species and varieties, as Darwin pointed out a good many years ago, and there is often difference of opinion as to whether a group should rank as a variety or as a distinct species.

Different species that closely resemble one another are classed as members of a larger group called a *genus;* a genus therefore is a group of similar species. Now it is convenient to give every species a name just as it is convenient to give every human being a name. The name of each species of animal or plant commonly consists of two words, the first designating the genus, the second the species. The Rocky Mountain locust is named *Melanoplus spretus*, Melanoplus being the genus and spretus indicating a particular species of that genus. The red-legged grasshopper is closely related to this species and is therefore placed in the same genus but given a different species name, *Melanoplus femur-rubrum*. The various species of bears mentioned are members of the genus Ursus which is the Latin name for bear; the grizzly bear is called *Ursus horribilis;* the common black bear, *Ursus americanus;* and the European brown bear, *Ursus arctos*. The naming of species of animals is very much like naming different people John Jones, Peter Jones and Mary Jones to indicate that they are individuals of the Jones family.

Just as similar species are grouped into genera so are allied genera united into a larger group, the *family*. Melanoplus, Dissosteira or the genus of the roadside grasshopper, and other related genera are grouped into the family Acridiidæ which includes the short-horned grasshoppers. Similarly various families such as the long-horned grasshoppers and katydids which constitute the family Locustidæ; the cricket family, Grillidæ; the cockroach family, Blattidæ, etc., are grouped into a larger division, or *order*, called the Orthoptera (*orthos*, straight, and *pteron* wing). The order Orthoptera, together with other orders such as the Coleoptera or beetles, the Diptera or flies, etc., constitute the *class* of Insecta. And the Insecta together with other classes such as the Crustacea (crayfish, crabs and their allies) and the Arachnida (spiders, scorpions, mites, etc.) and some others are united to form a still larger division called *phylum*, which is the largest subdivision of the animal kingdom. The vast assemblage of animals may be compared to a tree whose largest branches correspond to the phyla, the secondary branches representing the classes, the branches of these the orders, and so on, the leaves at the tips of the branches representing the species.

As it is often desirable to determine the group to which a particular animal belongs, guides to the proper group are often given in the form of a key. In using the following key to the families of the orthoptera ascertain first if the insect falls under the group A. If not try AA. If it goes in the latter division see whether it falls under B or BB. If it comes under BB, then find whether it belongs in C or CC. After finding the family, the genus and the species to which the insect belongs may be determined in a similar manner. As the genera and species of Orthoptera are so numerous, it is necessary to refer the students to more advanced works for this infomation.

Key to the Common Families of the Orthoptera

A. Posterior legs larger than the others and fitted for leaping.
   B. Antennæ much shorter than the body. Organ of hearing, when present, on the base of the first abdominal segment........ ....Acridiidæ, or short-horned grasshoppers.
   BB. Antennæ longer than the body. Auditory organs generally in the base of the tibia of the first pair of legs. Ovipositor generally long.
     C. Tarsi four-jointed, ovipositor flattened........Locustidæ, or long-horned grasshoppers.
     CC. Tarsi three-jointed. Ovipositor very slender and somewhat enlarged at the tip................................Grillidæ, or crickets.
AA. Posterior legs not much longer than the others and fitted for walking or running.
   B. Body rather short and broad, flattened; head partly inserted in the broad prothorax.........................Blattidæ, or cockroaches.
   BB. Body elongated.
     C. Fore legs large, spiny, fitted for grasping prey, prothorax long. ....Mantidæ.
     CC. Fore legs not large, prothorax short...............Phasmidæ.

CHAPTER III

THE LEPIDOPTERA OR BUTTERFLIES AND MOTHS

The butterflies and moths constitute the large order Lepidoptera, or scaly-winged insects. They have sucking mouth parts and undergo a complete metamorphosis. We shall select as a type of this order the common and widely distributed cabbage butterfly which is so frequently seen in our gardens. This butterfly was introduced from

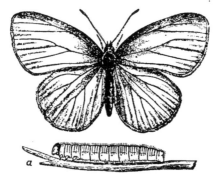

FIG. 14.—The cabbage butterfly. *a*, larva. (After Packard.)

Europe in about 1860 and, as it produces three or more broods a year, has now spread over most of the United States wherever there is a cabbage patch to afford food for its larvæ. The whitish scales which cover the wings and part of the body are beautifully sculptured objects, shaped much like a paddle with a very short handle. They are modified hairs and various intermediate gradations between hairs and scales may be found on different parts of the butterfly. The sucking tube which usually lies coiled up like a watch spring under the head may be straightened out into a long "tongue" by which the butterfly sucks the nectar from flowers. The sucker is formed of the parts of two modified maxillæ which have been enormously lengthened and closely fitted together. Their inner faces are concave so that a tube is formed when the two parts are applied. The mandibles in the butterflies are either represented by minute rudiments or absent entirely. There is little left of the labium except the two labial palps which project in front of the head.

The butterfly lays its eggs upon the leaves of the cabbage or some other related plant. Watch the butterflies as they flit about in a cabbage patch and you will probably see them alighting for a short time upon the leaves. Note carefully the spot where the butterfly rested; look at it with a hand lens and you may find a small oblong egg stuck by one end to the leaf. The egg soon hatches into a small green larva which is colored so nearly like the green leaves that it is often difficult to detect. The larva has mouth parts fitted for chewing, much like those of the grasshopper, the mandibles being particularly strong. There are three pairs of legs on the thorax; and on certain segments of the abdomen there are short, stubby legs called *pro-legs*, whose ends are furnished with minute hooks which aid the caterpillar in maintaining its hold on the surface of a leaf. The larva sheds its skin a number of times during its growth, and at its last molt passes into the *pupa* stage. It does not spin a cocoon as is done by many moth larvæ, but the pupa is fastened to some object by a thread which passes around the thorax. At the tip of the abdomen of the pupa there is an organ, the *cremaster*, which is furnished with hooklets for attaching to a small pad of silk which the larva spins just before transforming into the pupa.

In the pupa the sheaths, or cases of many organs of the butterfly, may be made out; the large wing cases lie at the sides of the body; ventral to these are the cases of the antennæ which show indications of numerous segments; next to the antenna cases are the cases for the first and second legs, those of the third pair being overlapped by other parts. Finally, in the middle line is the case for the tongue. Aside from being able to move the abdomen when it is irritated, the pupa is compelled to lead a stationary life. But while outwardly quiet the pupa is

Fig. 15.—The painted lady butterfly, *Vanessa atalanta*. (From photo by Essig.)

undergoing a rapid and extensive transformation of its organs. Old organs are being torn down, new ones are being built up, and extensive changes of form are taking place in other parts. So great are these changes that the pupa may be compared to a ship which has to be laid up for repairs. It is the workshop in which the caterpillar is being made over into the very different form of a butterfly. Wings are grown; large compound eyes replace the simple ones of the caterpillar; the biting and chewing mouth parts are modified into the elongated nectar-sucking proboscis; marked changes take place in the digestive organs which fit them for the very different diet of the butterfly; and changes equally great occur in many other parts of the body.

One of our commonest and most striking butterflies is the monarch, or milkweed butterfly. It has reddish-brown wings with black veins and a dark border with whitish spots. It is one of the few butterflies that are migratory; it frequently travels southward on the approach of winter, in large flocks. The larvæ live upon milkweed and may be recognized by their conspicuous black and yellow stripes surrounding the body. The pupa is green and hangs suspended by its cremaster. The pupa stage lasts about two weeks, the species passing the winter as an imago.

Closely resembling the monarch in the color of its wings is another butterfly of somewhat smaller size, called the viceroy, *Basilarchia archippus*. It can be most readily distinguished from the monarch by a black bar across the hind wings. The viceroy is not at all closely related to the monarch; the likeness is merely a superficial resemblance in color. As the monarch butterfly is particularly distasteful to birds it is therefore seldom troubled by them. The viceroy is commonly supposed to derive more or less protection from its resemblance to the monarch, since the birds would readily mistake it for the distasteful species. Such protective resemblance of one species to another is called *mimicry*. It is a curious fact that most of the other species of Basilarchia are colored very differently from the viceroy. There are a great many cases among butterflies in which a species may resemble in a most striking manner distasteful species of a quite unrelated group. At the same time these "mimicking" species may depart in an equally striking manner from the forms to which they are very closely allied in points of structure and life history. The

FIG. 16.—Mimicry of the monarch butterfly (upper figure) by the viceroy (lower figure). (After Lutz.)

resemblance to the distasteful species is superficial and due to color and outline of wings, while in many other less conspicuous features of structure they are similar to the members of their own genus or family. Similar cases of mimicry have been described in various other animals, but nowhere are the resemblances so numerous or so striking as in the butterflies, especially those of South America and Africa. A striking degree of protective resemblance is shown by many moths which are colored so as to be scarcely distinguishable when resting on the bark of trees.

FIG. 17.—Cecropia moth. (After Lutz.)

The butterflies, on the other hand, are usually colored so as to make them particularly conspicuous.

The moths constitute a very extensive group of the most varied sizes and colors. Generally the wings when at rest are horizontal or held folded over the abdomen, often sloping downward on either side. Usually the moths fly at night or in the evening, while the butterflies are lovers of the sunshine. Some of the largest and most conspicuous of our species belong to the giant silk-worm moths. One of these is *Samia cecropia*. The larva lives on the leaves of several kinds of trees; preparatory to going into the pupa stage it spins a cocoon of silk which is secreted by a pair of large glands opening upon the small lower lip. The cocoons are attached by one side to a twig and are formed of very tough material which is admirably adapted to keep out cold and moisture. The pupa passes the winter within this cocoon and the mature moth emerges in the spring. With no biting mouth parts the moth would be utterly unable to get out of its tough envelope, were there not left at one end an opening filled only with loose webby material through which it can push its way. A related species spins its cocoon against the side of a leaf so that the leaf becomes partially wrapped around it; and, as if to guard against its cocoon falling off the tree when the leaves are shed in the fall, the larva spins along the leaf stem a number of threads of silk connecting leaf and cocoon with the twig.

Perhaps the most beautiful of our moths is the large luna moth, easily recognized by its pale green color and by the swallow tails on its hind wings. Its larva feeds on the leaves of the hickory, walnut and other trees and forms a cocoon in which leaves are interwoven with the silk.

The moth most valuable to man is undoubtedly the silk-worm moth, *Bombyx mori*, which is a native of China where silk culture has been carried on for many centuries. The larvæ preferably feed upon the leaves of the mulberry although they will eat the common osage orange and a few other plants. The white or yellowish cocoon which the larva spins is constructed of a single thread which is generally over 1000 feet long. This thread is wound off on reels by the silk grower and then put through various processes of preparation according to the kind of silk product which it is desired to make. The moth makes its escape from the cocoon after moistening the end with a

FIG. 18.—Silk-worm moth eggs and cocoons.

secretion which softens the cementing substance between the fibers. It is a very sluggish sort of creature, scarcely able to fly, and takes no food.

FIG. 19.—A sphynx moth. (After Lutz.)

Toward evening one often sees the sphinx moths or hawk moths, (Sphingidæ) flying about in search of flowers out of which they suck nectar with their very long proboscis. Some are called humming-bird moths on account of the resemblance of their flight to the flight of the humming bird. One of the largest and best known of the numerous species of this family is the tomato-worm moth. It has a large, smooth, green larva with oblique white mark-

FIG. 20.—Tobacco worm. This gives rise to a sphynx moth. (After Howard.)

ings on the side and a curved horn at the posterior end of the body. It feeds on tomato and potato vines and upon tobacco.

The very large group of owlet moths or Noctuidæ include many of the most injurious species. The army worms which attack corn and grain, the cut worms of our gardens, the cotton boll worm which is estimated to cause over $2,000,000 damage a year to the cotton growers, and many other species which live upon the kind of vegetation that man happens to be interested in belong to this family. Belonging to a related family is the gipsy moth, *Porthetria dispar*, which was brought from Europe into Massachusetts in 1868. It spread with such rapidity and its larvæ did so much damage to shade trees and forest trees that the state organized a systematic attempt to exterminate it, spending over $1,000,000 in spraying, destroying the eggs, and in other methods of warfare. Meanwhile the gipsy moth thrives.

FIG. 21.—Imported currant-borer, with larva, *l*, and pupa, *p*.

Every boy who has eaten apples has doubtless come across the larvæ of the codling moth or else the evidences of its destructive activity upon his apple, although he may not have known that the offending "worm" is the larva of a small, dull-colored moth that lays its eggs at the blossom end of the developing fruit. The young larvæ eat their way toward the center of the apple, and when full grown they gnaw a hole to the surface and escape;

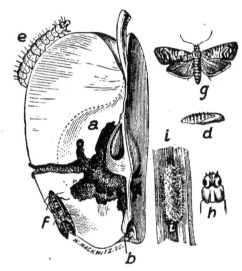

FIG. 22.—Codling moth, *a*, apple showing work of larva which enters at the blossom end *b*, where the egg is laid and finally eats its way to the outside; *e*, larva; *d*, pupa; *i*, cocoon; *f* and *g*, mature moths. (After Riley.)

then they crawl into some protected nook to pass through the pupa stage. The moths emerge in about two weeks and deposit eggs in other apples. The second crop of larvæ usually pass the winter in the apples and come out as moths the next spring. By spraying the young fruit with insecticides, and putting bands of cloth around the trees so as to catch and destroy the first brood of larvæ as they travel down to pupate in the ground, and by destroying the windfall apples, it is possible to reduce very greatly the ill effects of this pest.

The clothes moths may not only destroy what the silkworm moth has produced but they will also attack furs and all sorts of woolen cloth. There are several species of clothes moths all of which are of small size and similar appearance. They lay their eggs preferably in woolen garments, carpets, blankets, etc., the larvæ feeding upon the wool. There may be more than one generation a year. The damage done is entirely the work of the larvæ since

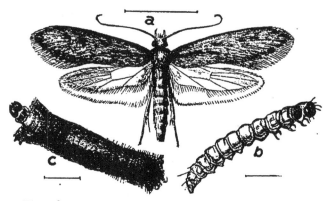

FIG 23.—The clothes moth. *a*, adult; *b*, larva; *c*, larva in case—enlarged. (From Riley.)

the moths themselves only deposit their eggs on the cloth. The moths are most active at night and are frequently attracted to lights. In many places persistent vigilance is required to keep these unobtrusive creatures from doing much damage. Clothing that is packed away should first be well brushed and exposed to light. If pasteboard boxes are used they should be sealed by pasting paper around the edges of the cover. If it is suspected that eggs or larvæ may still be in the clothing it should be kept in a tight receptacle together with about half a cup of bisulphide of carbon for twenty-four hours. The bisulphide will rapidly evaporate and penetrate all parts of the clothing and kill the eggs or larvæ if any are present.

CHAPTER IV

THE HEMIPTERA, OR THE BUGS AND THEIR ALLIES

By many unsophisticated people the term bug is applied to almost any sort of insect and even to a number of creatures which are not insects at all. In its stricter sense the word is used to designate any member of a certain division of the order Hemiptera. The Hemiptera in general include insects with a gradual or incomplete metamorphosis and sucking mouth parts. In typical members of the division Heteroptera, or true bugs, the anterior wings are thickened at the base, and the terminal portions are more or less membranous. The second pair of wings which are membranous are folded under the first pair and constitute the chief organs of flight.

FIG. 24.—The squash bug, *Anasa tristis*.

A good example of a true bug is furnished by the well-known squash bug of our gardens. The sucking organ which the bug uses to pierce and suck out the juices of plants consists of an elongated under lip, or labium, which is converted into a tube or sheath enclosing two pairs of very slender and sharp piercing organs which are commonly held to represent the highly modified mandibles and first maxillæ. The squash bug is protected from its enemies by means of stink glands which open on the side of the thorax and secrete an ill-smelling substance which is poured out when the insect is disturbed.

One of the most injurious of the bugs is the chinch

bug which sucks the juices of corn, wheat and other grains. Kellogg in describing the damage done by this bug in the Mississippi valley states that he has "seen great corn fields in this valley ruined in less than a week, the little black and white bugs massing in such numbers on the growing corn that the stalk and bases of the leaves were wholly concealed by the covering of bugs." The United States Entomologist estimated that the annual losses caused by the chinch bug amount to $20,000,000 a year.

While numerous species of bugs are destructive to vegetation there are many which prey upon other insects. Among these are the "assassin bugs," and the celebrated "kissing bug" which occasionally inflicts very painful bites upon human beings. The cone nose, or "big bedbug" occasionally attacks man also, but a more familiar and widespread pest is the ordinary bedbug of human dwellings. While these disagreeable insects possess but very small functionless rudiments of wings they can run with remarkable quickness. During the day they lie concealed in cracks and crevices, but at night they scurry about in search of their sleeping human victims at whose expense they gorge their bodies with blood. They breed with remarkable rapidity, and wherever they make their appearance it is therefore advisable to wage war upon them with the greatest vigor. For this purpose a saturated solution of corrosive sublimate in alcohol applied to the crevices where they lie concealed is an efficient remedy, although one which should be used with care as it is very poisonous. Besides being disagreeable on account of their bites and offensive odor, bedbugs are the means of transmitting certain diseases from one person to another. A disease common in India, and relapsing fever which sometimes occurs in the United States are transmitted by these insects.

There are many species of bugs that live in or on the water, and these forms show many remarkable and interesting adaptations to aquatic life. The large electric-light bugs, which are so frequently attracted by lights at night, live normally in the water where they prey upon small fishes and other aquatic organisms. They are fierce fellows and inflict severely painful bites. The slender water scorpion, which superficially resembles a walking-stick in the general form of its body, commonly lives near the banks of ponds and streams with the tip of its long posterior breathing tube protruded at the surface of the water. By its fore legs which resemble somewhat those of the praying mantis the water scorpion catches small creatures that come near, and holds them while it sucks their blood.

FIG. 25.—A water scorpion, Ranatra.

FIG. 26.—A back swimmer, *Notonecta glauca*.

Among the most interesting and easily studied of the aquatic Hemiptera are the back swimmers (Notonecta) which have the peculiarity of swimming with the dorsal surface downward. These insects commonly hang obliquely downward with the tip of the abdomen at the surface of the water to give them access to air. Their backs are light colored, a circumstance which renders them more nearly invisible to animals below the surface. The posterior legs are especially modified for swimming, being long, flattened and furnished with marginal hairs. When swimming below the surface the backswimmer carries quite a supply of air which gives it a silvery appearance where the light is reflected from the bubbles. Backswimmers live by sucking out the blood of small animals of various kinds, and they may attack insects larger than themselves. They are readily kept in aquaria, but too many of them must not be put together as they will sometimes attack and devour their own kind. Like many other aquatic Hemiptera, backswimmers are strongly attracted by light, and may be made to follow a light about in any direction. Similar in habits to the backswimmers are the water boatmen which also get their supply of air by placing the tip of the abdomen at the surface of the water.

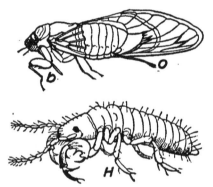

Fig. 27.—The periodical cicada. Upper figure, adult female; *b*, beak; *o*, ovipositor. Lower figure, young larva enlarged. (After Riley.)

While most of the aquatic Hemiptera live within the water, the water striders of pond skaters are found upon the surface, being supported by the surface film which they are not heavy enough to break through. Insects which fall upon the surface are quickly attacked by these watchful rovers.

In the sub-order Homoptera the fore wings when present are membranous throughout. One of the largest of this group is the harvest fly or cicada, whose prolonged and uniform note is often heard during the hot days of late summer. Another member of the same genus is the seventeen-year cicada or locust which has the longest period of metamorphosis of any known insect. After seventeen years of life spent in burrowing in the soil, feeding upon roots and other vegetable material, these cicadas make their appearance, often in great numbers, in the spring or early summer. The history of various broods in different parts of the country is known, so that it can be predicted, years in advance, when they will appear in any particular part of the country. The cicadas of the southern states may complete their metamorphosis in thirteen years.

Many of the smaller Homoptera such as the little green leaf hoppers are quite destructive to various species of

FIG. 28.—The bean aphis, *Aphis rumicis* Linn. Winged and wingless females with enlarged antennæ of the same. Greatly enlarged. (After Essig.)

plants. Some of the most injurious, as well as in some respects the most interesting members of the group are the aphids or plant lice. Many species do considerable injury by sucking the juices of plants, as they multiply with such remarkable rapidity that a plant may soon be literally covered by the descendants of a single individual. Aphids have the peculiarity of producing young from eggs which are not fertilized, a process known as *parthenogenesis*, a word meaning virgin reproduction. Usually several generations are produced in this way. These consist generally of wingless females, but at times winged females occur which may fly to another plant and give rise to a new colony. After a number of parthenogenetic generations, and generally upon the approach of cold weather in the fall, broods consisting of both males and females appear, constituting the so-called sexual generation. The females of this generation produce eggs of unusually large size which require to be fertilized before they develop. These eggs remain over winter and hatch out in the following spring into females which start a new series of parthenogenetic generations. Aphids are frequently attended by ants which imbibe a sweet liquid called honey dew which comes from the aphid's abdomen.

One of the most injurious of the aphid family is the Phylloxera which attacks the grape vine. In France especially, enormous injury has been done by this insect. It attacks both the leaves and the roots, producing peculiar galls in each, and causing the deterioration and often the death of the vine. Comparative immunity from these insects has been secured by grafting French vines upon vines native to America. The aphis which feeds upon roots of corn commonly lives in underground galleries of ants. The ants uncover the roots in their burrowing, carry the aphids to them, gather their eggs and carry them to places of safety; in return for these services the ants feed upon the sweet fluid derived from their adopted companions.

The most degenerate of all of the Hemiptera are the scale insects or Coccidæ. In a typical scale bug, such as the San Jose scale, or the apple-tree bark louse, the female is attached to a particular spot on a leaf or twig where she sucks in sap through her slender beak. In the scale bug, legs, wings, antennæ and eyes are lacking, and the body is covered by a scale, which is formed by a secretion from certain glands. The female lays large numbers of eggs which hatch into active young furnished with six legs, antennæ, and various special organs not found in the mature female. The young crawl about for a while, and soon become attached to one spot where they undergo a series of molts and change greatly in structure. At first the male and the female scale insects are very much alike, but after the first molt the male scale may be distinguished by its smaller size and narrower shape. While the female remains attached and becomes more degenerate with age, the male scale insect, after passing through a pupal stage, transforms into a small, graceful, winged in-

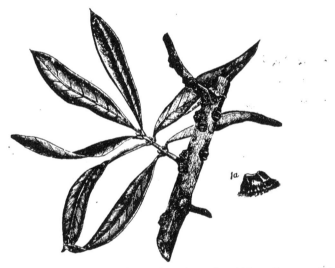

FIG. 29.—Black olive scale. *1a*, scale enlarged. (After Comstock.)

sect which flies from one plant to another. It has only one pair of functional wings, and owing to its imperfectly developed mouth parts it is incapable of taking in food; after fertilizing the female it soon dies. Males are in general more rarely seen than the females, and in some species of scale bugs they have never been discovered, the female probably reproducing exclusively by parthenogenesis.

The San Jose scale does a large amount of injury to fruit trees. It is a particularly bad pest in California, but despite efforts to check its spread it has become scattered through most of the United States. An allied species is a destructive enemy of orange and lemon trees. In some species the female retains the power of locomotion throughout life and possesses eyes, antennæ and legs. One of the less degenerate scale bugs is the cottony-cushion scale which secretes a cottony mass of fibers within which it deposits its eggs. This insect was introduced into California from Australia, and spread with such great rapidity that it threatened to exterminate the orange groves. An entomologist, Mr. Koebele, was commissioned to search for the enemies of the cottony-cushion scale in its native country and to import any species which might prove a means of checking the alarming spread of this pest. This search resulted in the importation of a beetle, *Novius (Vedalia) cardinalis*, which thrived and multiplied to such a degree that it effectively exterminated nearly all of the scale bugs of this species. Other scale bugs are combatted by spraying trees with kerosene emulsions and other insecticides, and by covering the trees with tents in which poisonous gases are generated in sufficient quantity to prove fatal to the insects without severely injuring the trees. All of these measures are expensive, but they are less costly than the damage done by the insects. Directly or indirectly the scale insects entail a loss of many millions a year. On the other hand there are a few species which are of economic value, such as the cochineal insect which feeds upon cactus in Mexico and

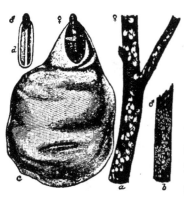

FIG. 30.—Scurfy bark-louse. *d*, male; *c*, female. (After Howard.)

yields us the coloring matter, cochineal; and the tropical *Carteria lacca* from which we derive shellac.

Lastly we must mention what are in some respects the most disagreeable of the Hemiptera, the lice, many species of which infest the lower animals and a few of which are peculiar to man. Fortunately they are not so prevalent as formerly, as they are not tolerated very long by people of the present average standards of cleanliness.

The Hemiptera are remarkable for their great diversity of form and habit. Active plant feeders, fierce assassins of other insects, degenerate and almost shapeless parasites of plants, denizens of the water, skaters on the surface film of ponds and streams, and the despised parasites of the human body, they nevertheless show certain features in common which indicate their kinship.

CHAPTER V

THE DIPTERA OR FLIES

The true flies or Diptera are, as the name implies, two-winged insects, although they have what really corresponds to the hind pair of wings of other insects in the

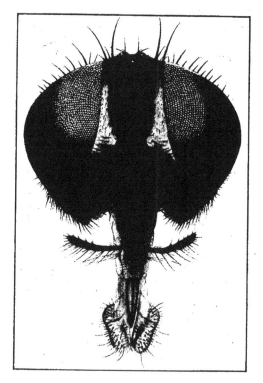

FIG. 31.—Head of housefly. (After Herms.)

small club-shaped *halteres* or balancers which are joined to the sides of the metathorax. All of the flies have sucking mouth parts which include in many cases slender piercing organs. All flies also pass through a complete metamorphosis with sharply differentiated stages. If familiarity always implied accurate knowledge it would be unnecessary to say anything concerning the common house fly (*Musca domestica*), but even entomologists have learned much that was new regarding this insect within the past few years. One of the most striking features of the fly's organization is the proboscis which has long been a favorite object with the amateur microscopist. Most of this structure consists of the labium or lower lip. Its end is flattened and divided into two lobes which can be folded together when not in use or spread apart and applied to a surface when the fly is sucking in food. The lower surface of the lobes is roughened so as to serve as a rasp or grater.

The antennæ of the fly are short and consist of three joints of which the last is much the largest. Examination of this joint with a microscope will reveal thousands of olfactory pits, the organs concerned with the sense of smell. This sense is very acute in flies, especially the blow flies and flesh flies. The foot of the fly, another favorite object of microscopists, shows in addition to two claws, a two-lobed flap which is furnished below with minute hairs at the end of which a sticky secretion is poured out, that enables the fly to walk up vertical surfaces and upon the ceiling.

House flies produce many broods a year. Their eggs are laid in horse manure and other refuse where they hatch in less than a day into white maggots. In about six days, the precise time depending upon temperature and food, the larvæ pass into the pupa stage, from which the mature insect emerges in about five days. Besides making themselves an inordinate nuisance in the house where it is apparently the height of their ambition to die in some article of food, house flies are a particularly dangerous means of spreading disease. The old contention that they are useful insects because they act as scavengers is a mischievous doctrine; as a matter of fact they possess no redeeming virtues. Since they wade about in all sorts of filth, they get germs of various diseases on their feet and may carry them to food. In many cases typhoid fever is definitely known to have been carried by flies, and it is quite probable that other diseases are carried in

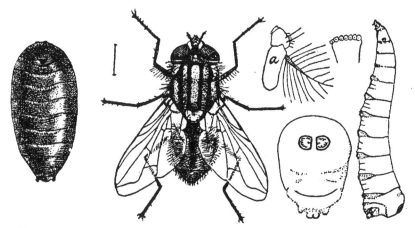

FIG. 32.—Common house fly, *Musca domestica*. Puparium at left; adult next; larva and enlarged parts at right—enlarged. *a*, antenna. (After Howard.)

the same way. To a certain extent flies may be gotten rid of by traps, fly poison and sticky paper, but as they are continually being bred in such enormous numbers it is best to check them by keeping covered the manure and other refuse in which they breed.

Closely related to the house fly are the blow flies and flesh flies whose larvæ live upon decaying flesh. The fly whose larva is known as the screw worm sometimes lays its eggs in wounds and in the nostrils of men and animals; the larvæ may devour the nasal membrane, and get into deeper passages connected with the nose and sometimes produce fatal results.

The larvæ of bot flies commonly inhabit the stomach or intestine of horses, cattle and sheep, attaching themselves to the mucous membranes of the walls and often causing great distress to the afflicted animal. The eggs of the horse bot fly are laid usually upon the hairs of the fore legs of the horse and gain access to the stomach when the horse licks itself. The larvæ remain in the alimentary

FIG. 33.—Larvæ of bot flies attached to the walls of the stomach of a horse. (After Osborn.)

canal during the fall and winter and pass out in the spring when they go through their pupa stage in the ground. The related warble flies cause much discomfort to cattle since the larvæ spend most of their life just beneath the skin. The full grown larva may reach a length of an inch. When ready to pupate it gnaws a hole through the skin of its helpless host and drops down and burrows in the ground. In this country the damage done to hides is estimated at $50,000,000 per year.

The family Tabanidæ, including horse flies, green heads and their allies, is formed mainly of blood suckers which pierce the skin with the sharp stylets of their proboscis. The larvæ generally live in the water or in damp situations; this is why the flies themselves are usually more abundant in low swampy regions.

Few of the Diptera are more annoying than the small, fiercely biting, black flies or buffalo gnats (Simulium) which occur in such numbers in certain parts of the United States and especially in Canada, as to make life almost intolerable at times for man and beast alike. The larvæ inhabit running water, attaching themselves to rocks and other objects by the posterior end of the body.

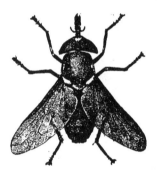

FIG. 34.—Black gadfly —enlarged. (After Howard.)

A large family of the Diptera, the Tachinidæ, make their living during their larval state as parasites within the bodies of other insects. The female fastens her eggs upon the skin of some insect and when the young grubs are hatched they bore into the body of their victim and proceed to devour its internal organs. In general the Tachinidæ deserve our gratitude because they afford a means of holding in check the ravages of destructive caterpillars, grasshoppers and other injurious insects.

Other scourges of the insect world are the robber flies, Asilidæ, which swoop down upon their prey, carrying it off and sucking its blood. We can no more than mention the hover flies, Syrphidæ, that have the curious habit of hovering over one spot during the summer days; the fuzzy bee flies, Bombyliidæ, that are often mistaken for bees; the long-legged, awkward crane flies, Tipulidæ; the beautiful and graceful midges; and the gall gnats, Cecidomyiidæ, which include the destructive Hessian fly which causes so great a damage to wheat fields.

It is desirable, however, to treat of the mosquitoes a little more fully, since these insects are perhaps the most important of all the insect enemies of man. Mosquitoes, like house flies, produce several broods a year. The females deposit their cigar-shaped eggs upon the surface of quiet water, some species placing them side by side with their pointed ends upward forming a sort of "raft" that floats on the water. The egg at ordinary summer temperature hatches in less than a day into a small, large-headed wriggler which is commonly seen attached at the surface of the water by its breathing tube situated near the posterior end of the body. The wrigglers live upon minute organisms and organic matter in the water, and they may often be found feeding near the bottom. They are under the necessity of coming to the surface for air at rather frequent intervals.

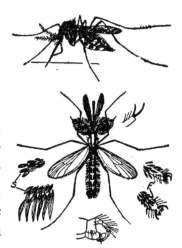

FIG. 35.—*Culex pungens.* Female above, male below. Feet, *f*, and scales, *s*, enlarged. (After Howard.)

After a number of molts the larva passes into a pupa stage which is peculiar in being active and in suspending itself, like the larva, at the surface of the water. Instead of having a posterior breathing tube, however, the pupa has a pair of such tubes attached to the greatly enlarged thorax. Projecting below the thorax may be seen the wing cases and leg cases which are closely fused to the body. The pupa takes no food and in a few days gives rise to the mature insect.

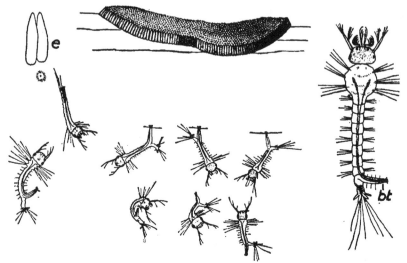

FIG. 36.—Egg raft and larvæ of the mosquito *Culex pungens*. *bt*, breathing tube; *e*, two eggs enlarged. (After Howard.)

Mosquito larvæ are commonly found in stagnant water which contains more or less decaying vegetable matter. They are not uncommon in rain-water barrels, cisterns

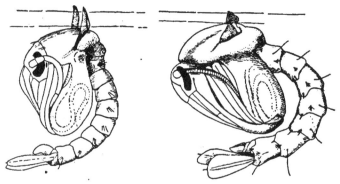

FIG. 37.—Pupa of Culex (at left) and Anopheles (at right)—greatly enlarged. (After Howard.)

and shallow wells, and may even live in cesspools. A few species live in salt marshes. The male mosquitoes may readily be distinguished from the females by their longer labial palpi, large, bushy antennæ and by having clasping organs at the posterior end of the body. In the great majority of species they are not blood suckers like the females, but content themselves with a vegetarian diet by sucking the juices of plants. For this reason and also because they are short lived, they are much less in evidence than the females.

It has been abundantly shown that the bite of mosquitoes of the genus Anopheles forms the sole means for the introduction of malaria into the human system. As will be more fully described in a later chapter the germs of

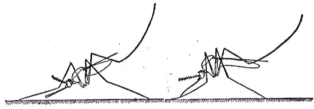

Fig. 38.—Anopheles at left, Culex at right in characteristic resting position.

malaria are introduced into the body with the fluid the mosquito injects from its salivary glands while it is sucking blood. In a similar way another much dreaded disease, yellow fever, is carried by mosquitoes of the genus Aedes (Stegomyia). By getting rid of disease-carrying mosquitoes, therefore, the liability of these diseases to spread becomes greatly reduced, and much attention has been given in recent years to devising efficient means of exterminating these insects. Draining swamps and marshes where mosquitoes breed is often resorted to, and where this is not feasible, covering the surface with a thin film of kerosene oil is often tried. Kerosene quickly kills the larvæ when they come to the surface to breathe, and while it may have to be put on the water more than once during the summer it has proven a very efficient remedy and one not very expensive when the area is not too great. Certain fishes, especially sun-fish, stickle back and minnows devour enormous numbers of mosquito larvæ, and the introduction of these fishes into waters where mosquitoes are in the habit of breeding has often nearly exterminated the larvæ. Careful attention to rain-water barrels and other stagnant water is also very desirable. In many places, especially where malaria or yellow fever occurs, a vigorous campaign against mosquitoes has greatly re-

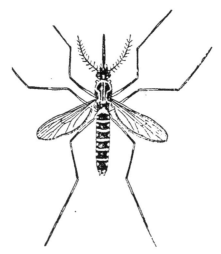

FIG. 39.—*Aedes fasciata*, the yellow fever mosquito. (After Howard.)

duced their numbers, and in most situations the trouble and expense of a mosquito crusade would be well repaid by the increased comfort to be enjoyed in the absence of these irritating pests.

THE FLEAS

It was formerly customary among entomologists to regard the fleas as Diptera which had lost their wings through disuse owing to their parasitic habits. But however they lost their wings, they are now generally ranked as a distinct order, the Siphonaptera, which may nevertheless have its closest relatives in the true flies. The fleas resemble most flies in havng mouth parts adapted for piercing and sucking, and their long and powerful legs give them such effective powers of rapid movement that wings would be almost superfluous. There are a great many species of fleas infesting different species of mammals and birds, but only a few attack man. The eggs are laid in hairs or among the feathers of the host and are usually shaken off. Hodge reports that "from a lady's dress on which a kitten had been fondled for a short time, fully a teaspoonful of flea's eggs were collected." The larvæ are slender, whitish grubs which feed upon dried animal and vegetable matter. There is a pupa stage of short duration, the whole period from egg to adult in the common cat and dog fleas being passed through in not more than two weeks.

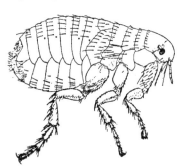

FIG. 40.—Female of flea, *Pulex irritans*, infesting man. (After Herms.)

Fleas, like mosquitoes, are disease carriers; one of the most dreaded diseases that afflicts mankind, the plague, which has carried off its hundreds of thousands in various epidemics that have swept over Europe and Asia, is carried by fleas. The disease attacks rats and squirrels as well as man, and wherever there are infected animals there is constant danger of an outbreak among human beings. The fleas coming from rats will readily carry the disease to man and from one man to another. Consequently when the plague was introduced into San Francisco a few years ago a crusade was made against the rats. Chinatown was subjected to the strictest search for these vermin and rats were trapped and poisoned in large numbers. Plague patients were carefully isolated, and the epidemic which had already made considerable progress, was fortunately held in check. Without the knowledge of the parts played by the flea and the rat in the dissemination of this terrible disease the country might have been swept over a scourge that would have claimed many thousands of victims.

CHAPTER VI

THE COLEOPTERA OR BEETLES

As the insects include many more species than any other class in the animal kingdom, so do the species of beetles outnumber those of any other order of insects. The order is not so diversified as the Hemiptera, notwithstanding its enormous size. In ninety-nine cases out of a hundred,

FIG. 41.—The California May beetle, *Lachnosterna errans* Lec. Adult and grubs, enlarged. (After Essig.)

one can detect a beetle at first glance, with perfect certainty. All of the beetles have biting mouth parts much like those of the Orthoptera, but they differ from the Orthoptera in undergoing a complete metamorphosis; that is, their life history falls into three well-defined stages: the active larva, the quiescent pupa, and the imago. The first pair of wings are modified into hard thick wing covers whose principal function it is to cover and protect the abdomen and the membranous second pair of wings which are the true organs of flight. Some of the most common and widely distributed of the Coleoptera are the June-bugs or May-beetles which often come buzzing in through the open windows of lighted rooms in the summer time. The larvæ are fat, whitish grubs which live upon the roots of grass and other plants, and frequently do con-

Fig. 42.—Life history of the Colorado potato beetle. *a*, eggs; *b*, larva; *c*, pupa; *d*, adult insect; *e*, wing case; *f*, leg.

siderable damage. After spending two or three years in the ground they pass through a relatively short pupa stage before emerging. There are many different species which are similar in form and habits.

The Colorado potato beetle is one of our most serious insect pests. Its original home was in the regions lying near the eastern slope of the Rocky Mountains and, as the cultivation of land extended from the east and the potato gradually traveled westward, the Colorado beetle found in this vegetable an acceptable food plant similar to the one it formerly preyed upon. The species then began to spread rapidly toward the east. In a few years it crossed the Mississippi, and in a few more years spread

Fig. 43.—The twelve-spotted cucumber beetle, *Diabrotica 12-punctata* Oliv. (After Essig.)

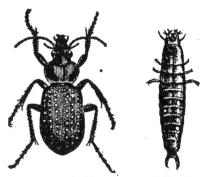

Fig. 44.—A "cut worm killer," *Calosoma calidum*, and its larva.

throughout all the middle and eastern states. The beetle is particularly destructive, since it feeds upon the leaves of the potato in the larval as well as in the imāgo stage. Its eggs are laid in masses on the leaves and they soon

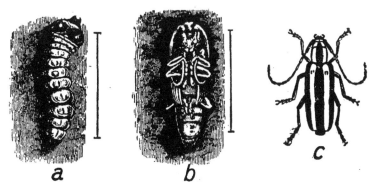

Fig. 45.—The round-headed apple-tree borer. *a*, larva; *b*, pupa; *c*, adult. (After Riley.)

hatch into reddish, fleshy larvæ which eat and grow rapidly. There are commonly two, and sometimes three generations a year. The mature insect passes the winter buried in the ground. The little, yellow, black-spotted cucumber beetle and the elm-leaf beetle which has caused so much damage to the elm trees in the eastern states are other destructive pests belonging to the same family as the potato bug.

The boring beetles, of which there are large numbers, include many destructive enemies of trees and shrubs. The larvæ, and in many cases the adults also, live upon the wood, forming tunnels which sometimes result in completely girdling the tree. The pupæ usually lie in the burrows and the mature insect gnaws its way to the outside.

FIG. 46.—Female of the Vedalia beetle.

FIG. 47.—A click beetle.

Among the beetles which are beneficial to man are to be counted the lady beetles which are of small size and usually hemispherical in outline. In both larval and adult stages they prey upon plant lice, scale bugs and other insects. The Novius previously mentioned which practically exterminated the cottony-cushion scale is a species of the lady beetle family.

Less useful, but on the whole beneficial beetles are the scavenger or carrion beetles which feed upon the decaying bodies of animals, which they can detect at a considerable distance through their remarkably acute sense of smell. There are numerous water beetles belonging to several distinct families. The Gyrinidæ, or whirling beetles, which are the analogues of the pond-skaters among the Hemiptera, are frequently to be seen darting about in groups at the surface of the water. Below the water live the Dystiscidæ which are active, predatory creatures, furnished with legs especially fitted for swimming. They come to the surface for air, and carry more or less air down with them under their wing cases. The larvæ have slender bodies and long, sharp jaws which are perforated, a device that enables them to suck the juices out of the forms on which they prey. Some other large water beetles (Hydrophilidæ) live upon decaying vegetable material as well as animal food.

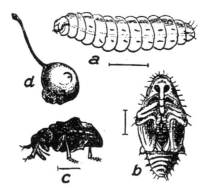

FIG. 48.—The Plum curculio, *Conotrachelus nenuphar.* *a*, larva; *b*, pupa; *c*, adult; *d*, weevil attacking young plums.

The snout beetles constitute a peculiarly specialized group, the head being prolonged into a snout which is often of considerable length. The very small mouth parts are situated at the end of this projection and the antennæ are borne on its sides. The group is a large one and includes forms known as curculios, and weevils, although the latter term is applied to some small beetles belonging to other families. The snout beetles are particularly destructive to fruits, grains, nuts, and to many other articles of food.

CHAPTER VII

THE HYMENOPTERA OR BEES, ANTS, WASPS AND THEIR ALLIES

The Hymenoptera, or membrane-winged insects, include the bees, ants, wasps and a host of less well-known forms, nearly all of which are characterized by having a combination of biting and sucking mouth parts. There are four membranous wings in the great majority of the Hymenoptera, and the females are provided either with a sting, or an ovipositor which is usually adapted for

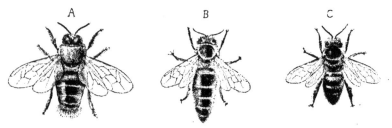

FIG. 49.—Honey bees. *A*, drone; *B*, queen; *C*, worker. (After Benton.)

inserting the eggs into the tissues of plants or the bodies of animals. This order includes the most highly developed of the insects, many of which are remarkable for the number and perfection of their instincts.

Among the higher Hymenoptera we find many instances of a highly developed social life which is well exemplified by our common hive bees. The hive bee community is composed of three kinds of individuals, the *queens*, the *drones*, and the *workers*. These three kinds or castes differ considerably in structure, and much more in their instinctive activities, for each caste has its particular set of duties to perform in the economy of the hive. The queen is larger than the drones or workers and has a relatively larger and more elongated abdomen. She is the fertile female, and her duties are limited to laying eggs in the cells made by the workers. Notwithstanding her name she is in no sense a ruler. The workers feed and

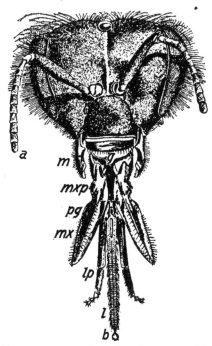

FIG. 50.—Head and tongue of worker bee magnified. *a*, antenna, or feeler; *m*, mandible; g, gum flap, or epipharynx; *mxp*, maxillary palpus; *pg*, paraglossa; *mx*, maxilla: *lp*, labial palpus; *l*, ligula, or tongue; *b*, bouton, or spoon of the same. (Reduced from Cheshire.)

protect her, and become uneasy and demoralized if she is taken away or dies, but she issues no orders to her faithful attendants. Each queen is actuated by a deadly hostility to every other queen, and if two queens happen to come together there is a fight to the death. As the young queen flies out of the hive she is followed by a number of drones with one of which she mates; she then returns and soon begins her duties of laying eggs. The drones do not gather honey and take no part in the household duties of the hive, but live idly upon the food collected by the workers. After the queen is fertilized and there is no more use for the drones, the workers fall upon them, sting them to death and drag out their dead bodies.

Although the worker bees are imperfect females, inasmuch as the development of the reproductive system has been arrested, there are few insects which possess so many wonderful adaptations of structure for their varied activities. Their mandibles are furnished with smooth edges, devoid of teeth, and especially fitted for moulding wax. The maxillæ and lower lip are modified to form a sucking tube, through which nectar is drawn from flowers. This nectar passes into the honey-sac or crop which is an enlargement of the alimentary canal just in front of the stomach; here it is converted into honey which is regurgitated into the cells of the comb.

The legs of the worker bee, in addition to their adaptation for walking and the pads and claws on the feet which fit them for climbing up either smooth or rough surfaces, show a number of interesting modifications for various other functions. One of these is the *antenna cleaner*. On the joint beyond the tibia, the metatarsus, is a semicircular notch which is lined by a very even row of spines, and at the lower end of the tibia there is a movable appendage which can be fitted over this notch so as to form a nearly circular space about the diameter of the antenna. When the antenna becomes covered with pollen or other material the bee throws its fore leg over it and pulls it through the antenna cleaner, thus stripping off the foreign material. The antenna cleaner is found also in the queen and drone and is common among Hymenoptera in general. The middle leg of the worker is fitted with a peculiar spine at the end of the tibia which is used for cleaning the wings and for prying off the masses of pollen which are carried on the hind legs. The third leg has a number of interesting devices for gathering and transporting pollen; on the outer side of the tibia there is a concavity bordered with stiff hairs, called the *pollen basket,* and if one watches bees that come in from the fields, these baskets may often be seen filled out with their yellow masses of pollen. How does the bee manage to get the pollen into the basket? This is done by means of the

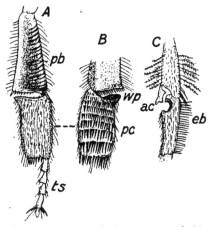

FIG. 51.—Legs of worker bee. *A*, lower part of third leg seen from the outside, *pb*, pollen basket; *ts*, tarsus; *B*, inner face of metatarsus showing the pollen comb *pc* and the wax pincers, *wp*; *C*, part of first leg showing the antenna cleaner, *ac*, and the eye brush, *eb*.

pollen combs on the inner side of the metatarsal joint. These combs consist of a number of rows of spines which are used to comb out the pollen entangled in the hairs of the bee's fuzzy body, and when they are filled the bee crosses its legs and rakes off the pollen from one comb into the basket of the opposite leg. Another ingenious device is shown in the *wax pincers* which are formed by the extended edges of the tibia and metatarsus at the place where they come together; these pincers are used for seizing the wax scales which are secreted by the wax glands. There is no pollen comb or basket and no wax pincers on the legs of either queen or drone.

The wax glands are found only in the worker. There are four pairs of these on the lower side of the abdomen. The wax is secreted in the form of thin scales which are seized by the pincers and passed forward to the jaws where they are mixed with saliva and kneaded into the proper consistency for making comb.

The sting which is present in the worker and queen but not in the drone is composed of two very fine and sharp "darts" which glide into a sheath at the posterior end of the body. There is a poison sac within the abdomen whose contents are forced into the wound in the process of stinging. The sting is really a modified ovipositor, a fact that explains its absence in the drone. Near the tip of the sting are some fine, recurved teeth which make the sting difficult to pull out when it has entered far into the flesh. In fact, the bees are frequently unable to pull out the sting and in freeing themselves certain organs of the abdomen are pulled away with the sting so that the insects pay the penalty of their boldness with their lives.

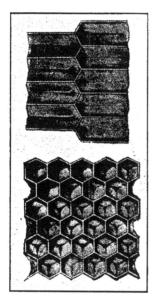

FIG. 52.—Worker cells of common honey bee (*Apis mellifera*); natural size. (After Benton.)

Gathering honey, keeping the hive clean, feeding the queen and young grubs, building the comb, and many other acts keep the worker bee well occupied during the period of its short life, which lasts in the summer only two or three months. The comb of the bee consists of six-sided cells on either side of a central plate of wax. The cells lie nearly horizontal inclining usually slightly upward. The whole arrangement of the cells is wonderfully adapted to afford the greatest amount of storage space with the least amount of material; a problem which the bees have solved as well as if they had a knowledge of geometry. The cells constructed for rearing drones are

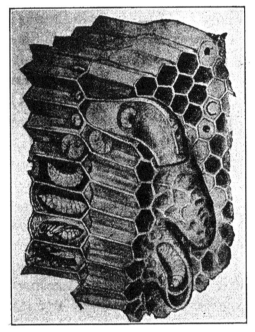

FIG. 53.—Comb showing eggs, larvæ, pupæ and queen cells. (After Benton.)

somewhat larger than the usual ones, but of the same shape; the queen cells, however, are much larger than the others and hang down at right angles to the rest and are usually roughened on the outer surface. Ordinary cells are used either for storing food or raising bees. Some of the cells are filled with a mixture of pollen and honey commonly called "bee bread" which is used in feeding the young larvæ or grubs.

Besides gathering honey or pollen, bees bring in a substance called *propolis* or "bee glue" which they obtain from the gums and resins of trees. This substance is used in stopping up cracks, for holding the combs in place, and for covering over offensive objects, such as dead slugs or other creatures which are too large to drag out of the hive. Bees are watchful nurses as well as good providers. The cells in which the queen lays eggs are supplied from time to time with honey and pollen as the young grubs require more food. Often one worker feeds another and all are ready and apparently eager to offer nutriment to the queen. Among bees care for their own interests means hostility to enemies, and the valiant worker is ready to sting any enemy that threatens the welfare of the community even though it involves the loss of her own life.

In the formation of new communities or swarms of bees the old queen followed by a swarm of workers issues from the hive. The swarm often settles on the bough of some tree from which the bees hang in a dense cluster. If left to themselves the bees may finally take up their abode in a hollow tree or some other protected situation, but the thrifty bee keeper usually transports the swarm to a new hive where they soon succeed in making themselves at home.

There are numerous species of bees which differ greatly in their mode of life. Many are solitary in their habits. These show no division into fertile and worker castes. The female in many species makes merely a shallow hole in the ground; this she stores with honey and pollen upon which she lays an egg; then she fills up the hole with dirt and leaves it. The larva, after devouring the stored up food, pupates and then emerges as a perfect insect. Some bees remain in the nest and care for the young more or less constantly, and thus form a family group. From the simple family there are various gradations to the primitive social community such as that formed by the common bumble bees. Here the queen, which is the only member to survive the winter, starts a nest during the spring in a hole or some depression in the earth, which she often covers over by bits of moss or grass. Then she makes a few waxen cells, stores them with honey and pollen, and lays in them eggs which hatch into worker bees. The workers are of relatively small size, but otherwise in appearance they are very similar to the queen. They busy themselves with making new cells, storing them with honey and pollen, and feeding the young grubs. Later in the season queens and drones appear; the queens after becoming fertilized scatter, and those that survive the winter found new colonies in the following year.

FIG. 54.—Nest of the solitary burrowing bee, *Osmia*. *e*, egg; *p*, pollen.

The bumble bee community is not a permanent one, but the transitory product of a single season. In its household arrangements, as in many other respects, it is simple and crude compared with the social life of the hive bees; the wax cells are rounded capsules arranged in no very definite order, and there is only one kind of cell produced. The queen at first, as in the solitary bees, performs all the labors of making a nest and rearing young, and only later devotes herself exclusively to laying eggs. Division of labor is not carried very far and the castes are specialized only to a slight extent. There are other bees whose social life is more complex than that of the bumble bees; they form connecting stages between the latter and the hive bees whose social arrangements represent the culmination of a long life of evolution from the solitary species.

In the wasps, as in the bees, there are both solitary and social species. The solitary wasps commonly prey upon insects and spiders which they store in their nests as provision for their young larvæ. The habits of a number of species have been carefully studied by Dr. and Mrs.

FIG. 55.—Solitary wasp, Ammophila, stinging a caterpillar. (After Peckham.)

Peckham who have written a most interesting book upon the results of their investigations. Each species generally hunts a particular kind of prey and constructs a particular kind of nest. In many cases the prey is stung so as to paralyze it without destroying its life, thus providing the larvæ with a supply of fresh food. The digger wasp Psammophila makes a little hole in the earth, and then goes in search of a caterpillar which it proceeds to sting on the ventral side of the body near the nerve centers. Then the wasp flies across the fields with her burden to her inconspicuous little hole, and after dragging her prey in she lays an egg upon it and covers it over with earth. Often several caterpillars are put into the same hole and after the last one is disposed of, the wasp fills up the hole and leaves it. The mud daubers frequently build their nests on the sides of buildings; they make cells of mud and commonly store them with paralyzed spiders.

In the social wasps we have a worker caste consisting of sterile females which show few external differences, except in size, from the fertile females. The best known species are the yellow jackets and hornets. Both these forms build rather large nests of a paper-like substance which they make by chewing up wood. Out of this paper they construct remarkably neat and regular six-sided cells which are placed with their open ends hanging downward. There are commonly several tiers or stories of these cells one over the other, and the whole is surrounded with a paper envelope with a hole in the bottom. The large white-faced hornets attach the nest to the branch of a tree, and it is not safe to molest them as the hornets have a very irritable temper and can sting with considerable severity. The nests of some species of yellow jackets are commonly found under the ground while others may be attached to buildings or trees. These creatures are likewise very pugnacious, but not so formidable as the hornets.

Ants

One would not at first sight regard the ants as members of the order Hymenoptera because most of them are wingless, but study of their structure shows them to be related to the bees and wasps. It is only the workers, and these are by far the most numerous, that never develop wings; the males and fertile females are winged as they emerge from the pupa state, when they are commonly spoken of as "flying ants." Like the bees the ants fly out of the nest to mate; after the breeding period the males soon die and the females strip off their wings and spend the greater part of their life in the wingless state. Sir John

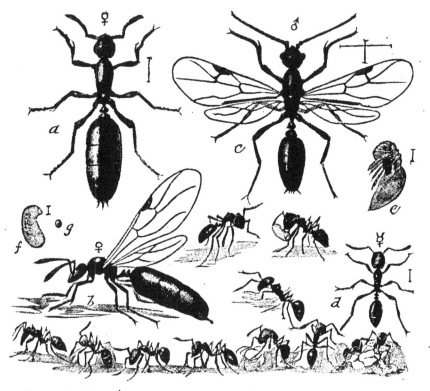

Fig. 56.—The little black ant (*Monomorium minutum*). *a*, female; *b*, same with wings; *c*, male; *d*, workers; *e*, pupa; *f*, larva; *g*, egg of worker—all enlarged. (After Marlatt.)

Lubbock records having kept a queen ant for nearly fifteen years.

In some species there are more than three castes, because the sterile females are differentiated into soldiers, workers proper, and more rarely other kinds of individuals. The soldiers are relatively large and have big heads and strong jaws; they generally take a prominent part in the defense of the community against its enemies.

Ants usually make nests by digging burrows or tunnels in the earth, and heaps of material are often accumulated around the openings of the burrows, or "ant hills" with which everyone is familiar. The large, black carpenter ants burrow into old stumps or rotten tree trunks. A few species make nests in the hollowed-out stems of plants. The nests are usually in places where it is moist and dark. Here the eggs are laid, the larvæ tended and fed by the workers, and the pupæ stored in suitable chambers. The pupæ of many species are enclosed in a cocoon spun by the larva. These cocoons are often erroneously called "ant eggs," the true eggs being very much smaller objects. Both eggs and pupæ are objects of much solicitude on the part of the worker ants; they carry them about from one chamber to another so as to keep them in a favorable situation, and when a nest is broken into the workers may be seen wildly rushing about with pupæ or egg-masses in their jaws in the effort to save them from destruction.

Most students of ant life agree that ants have a power of communication by means of striking one another with their antennæ and by making other signs whereby they may be warned of danger, or induced to follow a particular ant to obtain food. If one ant discovers a bit of sugar it is not long until a train of other ants is trooping to the spot. The ant community is closely bound together in its common interests; the members work industriously for the common good, and are ready to engage in fierce struggles for the defense of their community. Stir up an ant hill and you will see with what vigor and zeal the ants rush out to attack the offender. War in many species is an almost chronic condition. This is especially true of the slave-making species which regularly make excursions against other communities, and after killing off or driving away the inhabitants, seize the pupæ and carry them back to their own nests. Here when the young ants emerge they are adopted by their captors and take part in the household and other duties as if to the manner born. In some cases, as in the Amazon ant, *Polyergus rufescens*, the ants have come to be dependent upon their slaves for their subsistence. The Amazons have large curved jaws especially fitted for fighting, but they have lost the power of food gathering and would perish were they not fed by their faithful slaves. They have become so specialized as warriors that they have become useless for all other tasks. In addition to their slaves, ants sometimes harbor in their nests a motley assemblage of other creatures which are often spoken of as "guests" or commensals. We have already spoken of the aphids or "ant cows" from which the ants obtain a sweet juice. Ants also harbor many species of beetles including several blind forms; these are cleaned and fed by their hosts with as much care as is bestowed upon members of their own family. In return the ants obtain a secretion from these guests. In many cases the inmates of the nest seem to be simply tolerated without affording the ants any compensation for their board and lodging. There are known to be over one thousand species of insects which live for all or a part of their lives in the nests of ants, and many of them show very curious adaptations to this mode of life.

Ants are sometimes a source of considerable annoyance to man. The little red ants that come into houses and delight in getting into the sugar and other articles of food are often difficult to deal with on account of their small size. The best way to check them is to follow up their runway to the nest and then flood the insects with kerosene, gasoline or bisulphide of carbon. In the south and in California, the Argentine ant which was recently introduced from Argentina is proving a serious pest to the fruit growers, and there are several other species which which make themselves more or less of a nuisance.

The species of ants are very numerous and they are found in nearly all parts of the earth. •It would require a volume to treat of the peculiarities of these interesting insects, and we can no more than mention the remarkable honey ants, the harvesting ants, the leaf-cutting and fungus-growing ants, and the ferocious driver ants; all of these it would well repay the student to look up in larger works.

Besides the ants, bees and wasps, the Hymenoptera include a number of less well-known families. The ichneumons and their relatives lay their eggs on or in the bodies of insects or insect larvæ, and the young feed upon the tissues of their host, thereby proving of great value in checking various insect pests. The members of the family Cynipidæ are commonly known as the gall flies. When the eggs of these insects are deposited on or in the tissues of plants a peculiar growth of the vegetable tissue results, which is called a gall. The shapes of these galls are very characteristic; the gall caused by a certain kind of insect in a particular species of plant differs from the gall produced by another insect in the same plant and also from that produced from the same insect in a different kind of plant. Galls may be produced by members of various other families of insects, such as the aphids and certain flies. They represent an abnormal growth of plant tissue caused by the presence of some irritating material, but they are of use to the insects producing them because they afford both food and shelter for the young.

CHAPTER VIII

THE DRAGON FLIES, DAMSEL FLIES, MAY FLIES, STONE FLIES AND CADDIS FLIES

All of the insects described in this chapter spend their early or larval life in the water, and the adult insects are frequently seen near the water, although they may at times fly to a great distance from it. The dragon flies have four wings of similar shape and nearly equal size, which are held out horizontally when the insect is at rest. They have strong biting mouth parts and enormous compound eyes which cover over a large part of the surface of the head and enable the insect to see in almost all directions at once. Vision in the dragon flies is very acute, as it must be to enable them to catch the small flying insects which furnish their food. Their powers of flight are developed in proportion to their keenness of vision. Watch a dragon fly darting through the air in the hunt for prey, or better still attempt to catch one in a net, and you will appreciate the efficiency of its eyes and wings. Although dragon flies are called "snake feeders," "devil's darning needles," and other uncomplimentary names, and are associated, like the praying mantis, with a lot of foolish superstitions, they are entirely harmless, and indeed very useful creatures, since they devour many other insects, including a considerable number of mosquitoes. The eggs of dragon flies are laid in the water, and hatch out into dull-colored, inconspicuous, slow-moving larvæ, which prowl along the bottom after prey or lie in wait for it to come near. The larva is furnished with a peculiar organ for seizing prey, the so-called mask, which is really the labium, or second maxillæ. It is furnished with movable hooks at the end and is capable of being extended for a considerable distance in front of the head. If an unwary insect ventures too near, the mask is shot out with great rapidity and the insect pulled back to the mouth. When at rest the mask is folded up under the head.

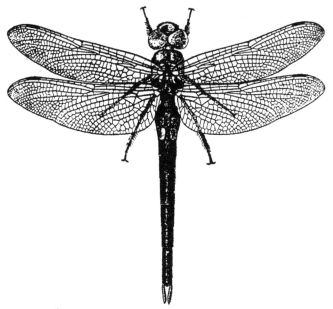

FIG. 57.—Dragon fly. (After Kennedy.)

After a series of molts, during which the rudiments of the wings are gradually developing, the larva becomes ready for the final change into the imago or mature insect. It then crawls out of the water, the skin splits down the back and the dragon fly pulls itself out of the old skin, shakes its wings dry and flies away. Upon a superficial view there is little resemblance between the alert and graceful movements of the one as it darts through the air with the greatest quickness and precision, and the sluggish stealthy prowling of the other in the bottom of a pond and stream.

The damsel flies have a slender body, broad head with widely separated eyes, and very delicate wings which are held longitudinally over the abdomen when at rest. The damsel flies are frequently conspicuous on account of their beautiful coloration, and they are generally found flitting about near the water. The larvæ usually have long flattened appendages at the end of the body called tracheal gills which are abundantly supplied with tracheal tubes for carrying air which is absorbed from the water.

The larvæ of the May flies resemble in many ways those of the dragon flies, but they may usually be distinguished by a number of tracheal gills attached to the sides of the abdomen, and by the long thread-like feelers at the end of the body. The gills are kept moving back and forth, thus keeping the water near them in constant circulation. The larvæ of some species of May flies live in the water for two, or even three years. They shed their skin many times, their wing buds becoming larger with successive molts. Finally they come to the surface, and the winged insect emerges from the nymphal skin. This process may occur when the larvæ is lying at the surface of the water, the old skin serving as a sort of raft or float which keeps the winged insect from getting wet. The life of the mature May fly is very short, some forms living but one or two days, hence the term Ephemeridæ which has reference to the ephemeral existence of the members of this family. Soon after the insects emerge they cast a very delicate skin, mate, and then lay their eggs in the water. They take

FIG. 58.—May-fly. (From Packard.)

no food in the imago state; in fact their mouth parts are so much reduced and atrophied that it would probably be impossible for them to do so if they made the attempt. The only function which the imago stage subserves is that of reproduction. May flies frequently appear in vast numbers in the vicinity of bodies of water and at night they are often attracted to lights, under which the dead accumulate in great heaps. Often the dead May flies drop into the water and are washed ashore in masses resembling large windrows.

The caddis flies are much better known in their larval than in their adult state. The larvæ are remarkable for surrounding themselves in a tubular case made of various materials which they carry about with them. Some species construct cases of sand, some employ irregular sticks of wood, while a few make their cases of bits of leaves which are cut out in a regular rectangular shape and fastened together at the edges in a most neat and orderly manner. Usually only the anterior part of the body is protruded from the case. The posterior part is soft and generally furnished with outgrowths which serve as gills, and at the tip of the abdomen there is a pair of hooks by means of which the worm holds on to its case. When the worms are removed from their cases they will readily construct new ones if given the proper materials. Caddis worms pass through a more or less quiescent or pupa stage after the close of their larval life. The larva closes up the end of its tube and transforms into a pupa, which lies within the old case. The mature caddis flies are somber colored, inconspicuous insects that are usually not much in evidence. They frequent places near the water in which they lay their eggs.

The stone flies are similar in their habits to the May flies. The nymphs generally live under stones in ponds

and streams and the adults usually fly ne
Another insect of somewhat similar habit
dobson fly, Corydalus, which reaches a
inches. The males have remarkably long
dibles. The large larva (dobson or hell
much prized as bait.

CHAPTER IX

THE MYRIAPODS AND ARACHNIDS

The Myriapoda which include the centipedes, millipeds and their allies constitute a group more or less closely related to the insects and from which the insects probably were developed. They have usually an elongated body with many segments and many pairs of legs. There is a well-defined head furnished with antennæ, mandibles and maxillæ, but there is no division of the body into thorax and abdomen as in insects, the various segments being remarkably similar in character except near the extreme end of the body.

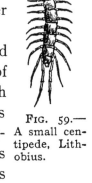

FIG. 59.—A small centipede, Lithobius.

The centipedes have a flattened body and are furnished with a strong, curved pair of jaws just behind the head at the tips of which opens the duct of a poison gland. These jaws are really the modified legs of the first segment of the body, and they serve as a means of injecting poison into the insects or worms on which the centipede commonly preys. The larger centipedes are capable of inflicting very painful bites upon man.

The millipeds or "thousand legged worms" are mostly cylindrical in form and have two pairs of legs attached to each ring in most of the segments of the body. There are no poison-bearing jaws as in the centipedes, but many species secure protection by means of stink glands which open along the sides of the body and pour out an evil-smelling secretion. The millipeds are mostly vegetable feeders, and some species are more or less destructive to crops.

The myriapods in general are lovers of the dark and are commonly found under rocks and logs. Some species are remarkable for the great elongation of the body and the numerous segments composing it. In Geophilus the segments may be over 170 in number. It may have been some such creature that induced Professor Ray Lankester to write:

> A centipede was happy quite
> Until a toad in fun
> Said, "Pray, which leg moves after which?"
> This raised her doubts to such a pitch,
> She fell exhausted in the ditch,
> Not knowing how to run.

FIG. 60.—Poison gland of spider, Nemesia, with duct and fang.

The Arachnida include scorpions, spiders, ticks, mites and a number of other forms which differ very greatly in size, structure and habits. We shall consider first the spiders as they are the most familiar and the most important. Spiders differ from insects in possessing eight legs and in having the head and thorax fused into a single piece. There are usually eight simple eyes, but no trace of antennæ. The head is furnished with a single pair of jaws which end in sharp curved hooks at the end of which is the opening of a large poison gland. Behind the jaws are the *maxillæ*, the bases of which are expanded into a sort of lip, the rest of the organ constituting the jointed *palp* or feeler. Spiders never chew their food; they simply suck out the juices of the prey which is held by the fangs. On the under side of the abdomen near the base are two slits which open into the pulmonary sacs which contain the organs of respiration. These organs consist of a series of flattened plates arranged like the leaves of a book and are therefore called the *lung books*. In addition to these organs spiders are furnished with tracheæ or air tubes similar to those of insects, but opening near the posterior end of the body. The sex opening occurs just behind the slits leading to the pulmonary sacs.

The organs which are the most distinctive of spiders are the spinning organs which produce the material of

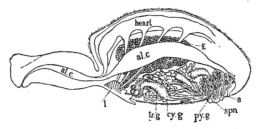

FIG. 61.—Section through the middle of an orb weaving spider. *al.c*, alimentary canal opening at *a*; *E*, eggs in ovary; *l*, lung book; *cy.g*, *py.g*, and *tr.g*, cylindrical, pyriform, and tree-like spinning glands respectively; *spn*. spinnerets. (After McCook.)

the spider's web. These consist of numerous glands located in the lower part of the abdomen; they open through the spinnerets of which there are usually three pairs near the posterior end of the body. There are in each spinneret a large number of small tubules each of which is connected with a duct from a spinning gland. The material of the web is at first soft and sticky. The spider, after attaching its web by placing its spinnerets against some object, draws out the soft material which rapidly hardens. As there are numerous spinning tubes the web of the spider, fine as it is, consists of a large number of strands. The web of spiders is used for many purposes. It is employed to make the cocoon with which the female surrounds the eggs, in making nests of various sorts, and in the construction of snares for the capture of prey. The snares may consist of irregular masses of web, broad funnels which lead into a tubular retreat, or, as in

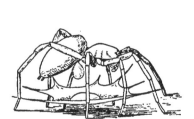

FIG. 62.—Female Drassus in the act of dropping eggs. (After Emerton.)

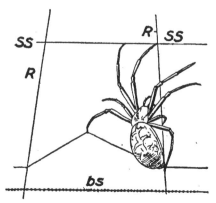

FIG. 63.—Orb weaving spider spinning. *bs*, beaded spiral of sticky web; *R*, rays; *SS*, smooth spiral. (After McCook.)

the orb weavers, of a beautiful circular orb of remarkable regularity and beauty of construction. The orb weaving spider commonly hangs downward in the center or hub

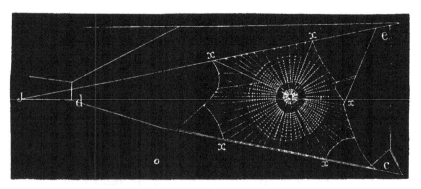

FIG. 64.—Diagram of an orb web. *je, dc, ec*, scaffolding of web; *x, x, x*, foundation lines to which the rays are attached. (After McCook.)

of the web with its legs spread out upon the radiating strands, ready to hasten to any part which is disturbed by the struggles of an entangled victim. In their efforts to overcome a large insect caught in their snares, many orb weavers make another use of their web in spinning a broad sheet of it around their captive until its struggles are effectually overcome. Then the spider sucks out its victim's blood and frequently ejects the carcass from its snare.

Another interesting use of web is in "ballooning" which is a common practice among young spiders. The spider when preparing for its journey through the air mounts upon a prominence and shoots out a quantity of web which is

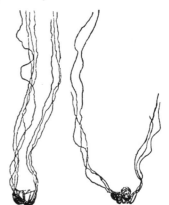

FIG. 65.—Attitude of aeronautic spider just before taking flight. (After McCook.)

FIG. 66.—Ballooning spiders floating in air by means of masses of web. The spider at the right is gathering in its web in order to descend. (After McCook.)

caught by the wind and pulled out further; finally the spider lets go its hold and is carried away with the floating mass of web that is wafted by the wind. Spiders may travel for miles in this way. When they wish to descend they haul in the web until they slowly sink to the ground.

There are many spiders which actively hunt for their

prey instead of spreading snares for it. Among these are the so-called running spiders and the jumping or hopping spiders. The latter especially are easily kept in confinement where one may readily observe their alertness, pugnacity, keenness of vision, their careful tactics in approaching their prey and many other entertaining peculiarities of behavior.

The adult males of spiders may be distinguished from the females by the enlarged terminal joint of the palps. The palps serve as the organs of sperm transfer at the time of mating. The males after having drawn the spermatic fluid into the enlarged end of the palp convey this material to the sex opening of the female and thus effect a fertilization of the eggs. Fertilization is often preceded by an elaborate courtship in which the males perform the most curious antics which are often considered to be the means of displaying their charms, the female being supposed to choose the most attractive male. In some spiders the males are many times smaller than the females, and courtship is attended with its dangers, since the female often pounces upon her small suitor and devours him without the least ceremony. Spiders in general are creatures with little sympathy in their composition. Their chief business in life is preying upon other creatures, and they have an attitude of hostility to almost everything that moves. They perform a valuable service to man in killing off millions of injurious insects.

There are very few species whose bite is at all dangerous or even painful. It is best not to be too familiar with the large tarantulas of the south and west, as their bite is very poisonous. There is a small round-bodied, black spider, Latrodectes, common in the south and west, which has a bad reputation, as many cases of severe poisoning and some fatal ones have resulted from its bites. As a rule spiders are not only harmless, but useful creatures.

The scorpions are arachnids which may very easily be recognized by their large pincers or claws, and their long jointed tail which ends in a sting. What corresponds to the mandibles of the spiders are small, short pincers, the large ones corresponding to the spider's maxillæ.

FIG. 67.—A scorpion, *Vejovis boreus*. (After Essig.)

The scorpions are predatory animals and live under rocks and in other protected situations, commonly in warm climates.

The ticks and mites form a more or less degenerate group which contains many parasites on both animals and plants. The ticks live upon the blood of various animals and not infrequently attack man. They have a very quiet way of boring their head into one's skin, and they are frequently not discovered until they have so gorged themselves with blood that they are many times their original size. After a full meal they may live considerably over a year without food. Ticks, like mosquitoes, are the means of spreading disease. Texas fever which sometimes exterminates many thousands of cattle is carried from one animal to another by ticks. Rocky Mountain fever, a disease of man prevailing in certain parts of the west, is also carried by ticks. A knowl-

FIG. 68.   FIG. 69.

FIG. 68.—Tick that produces the Rocky Mountain fever. (After Hunter and Bishopp.)

FIG. 69.—Female of Texas fever tick laying eggs. (After Hunter and Bishopp.)

edge of the mischievous rôle of ticks has resulted in greatly checking the spread of both these diseases.

The mites form a very large group of usually very small creatures. Some live in water, others in damp soil, others on plants, others attack animals. The chicken mite which causes so much discomfort to our poultry, is a common pest. The small cheese mites are a frequent nuisance, but perhaps the most acutely disagreeable of all are the very minute mites which produce the disease, now happily becoming more rare, known as the itch.

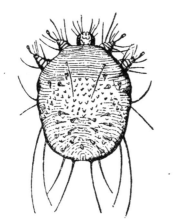 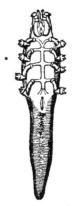

FIG. 70.—*Sarcoptes scabei*, female itch mite. (After Leuckart.)

FIG. 71.—*Demodex folliculorum*, follicle mite. (After Ludwig-Lennis.)

A degenerate form, Demodex, is a common parasite in the follicles of the human face and produces the appearance known as black heads.

CHAPTER X

THE CRAYFISH AND OTHER CRUSTACEA

It is a convenient and common practice to study the crayfish as a type of the Crustacea, especially ever since Professor Huxley so used it, and wrote his celebrated volume on this animal which can be heartily recommended to every student of biology. The body of the crayfish is covered by a chitinous external skeleton as in insects, but it is hardened by deposits of lime salts, except at the joints where it remains thin and flexible. The head and thorax are fused into one piece, the *cephalothorax*, which is covered dorsally and at the sides by a sort of fold called the *carapace*. Anteriorly the carapace is prolonged into the *rostrum* or *beak*. The part behind the cephalothorax is the *abdomen* and consists of seven freely movable segments or somites, the terminal one being called the *telson*.

The eyes of the crayfish are compound and situated on movable stalks. There are two pairs of antennæ; the first pair has two slender flagella, the outer one of which bears minute, club-shaped bodies which are organs of smell. The long second antennæ are mainly used as organs of touch; the first segment, however, contains the opening of the "*green gland*" or organ of excretion.

The crayfish is furnished with six pairs of mouth parts, the first of which, the *mandibles*, are stout organs well adapted for crushing food. The two following pairs are the *maxillæ*, and these are followed by three larger pairs of appendages called the *maxillipeds*. The first pair of legs are called *chelipeds* because they end in a pair of *chelæ* or pincers. There are small chelæ on the two following pairs, but the two posterior pairs of legs end in a simple claw. The large chelipeds are used as organs of defense and in the capture of prey. The other legs are all employed in walking, but the small chelipeds are used also in cleaning the body, and in picking up small bits of food and passing them forward to the mouth parts. The abdomen is capable of being curved downward and extended, the various segments being articulated by hinge joints at

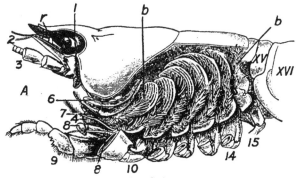

FIG. 72.—Crayfish with the side of the carapace cut away to show gills. *b*, branchiæ or gills; *r*, rostrum; *l*, eye; 2, first antenna; 3, second antenna; 4, mandible; 6, second maxilla; 7, 8, 9, maxillipeds; 10–14, bases of legs. (After Huxley.)

the sides. On each abdominal segment except the telson there is a pair of appendages (swimmerets). In the male the first two pairs are modified into organs for the transfer of the sperm cells; the other appendages are nearly alike in both sexes. The posterior pair is furnished with two expanded branches which, with the telson, form a tail fin used in swimming. When the crayfish is disturbed in the water it suddenly bends the abdomen downward and forward, thus causing the animal to dart quickly backward. In the female the small abdominal appendages are used for carrying the eggs which become attached to the hairs of these organs by a sticky secretion.

On either side of the thorax, in a space called the branchial chamber which is covered over by the sides of the carapace, are the breathing organs or *gills (branchiæ)*. The crayfish breathes the air which is dissolved in the water. A fresh supply of water is kept passing over the gills by the movements of a lobe on the second maxilla; the water is expelled from the anterior part of the gill cavity and passes into it from the sides under the edge of the carapace. If one places a little colored fluid in a dish containing a crayfish the fluid can be seen to be drawn into the chamber

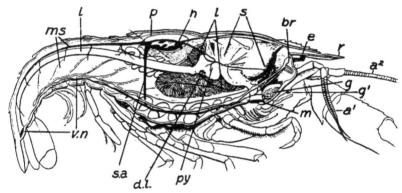

FIG. 73.—Anatomy of the crayfish. a^1, first antenna; a^2, second antenna; *br*, brain; *d.l.*, duct of liver; *e*, eye; *g*, green gland or excretory organ opening at g^1, *h*, heart; *i*, intestine, *l*; liver; *m*, mouth; *ms*, muscles; *p*, pericardium or sac surrounding the heart; *r*, rostrum; *s*, stomach; *sa*, sternal artery; *v.n.*, ventral nerve cord. (After Hatschek and Cori.)

and expelled in a stream in front of the body. The gills of the crayfish are feather-like structures consisting of a large number of filaments attached to a central stem or axis. Some of the gills are situated on the bases of the walking legs, others are attached to the sides of the body. They are to be regarded as complex and greatly branched out-pushings of the surface of the body in order to afford a great increase of surface exposed to the water; the walls of the filaments are very thin in order to facilitate the exchanges between the blood within and the surrounding medium.

While the crayfish often consumes vegetable matter as food it is generally carnivorous in its habits, living upon worms and various other living creatures that it may catch, and often devouring dead and partly decayed flesh. The food after being passed to the mouth parts is chewed mainly by the mandibles and is swallowed through a short tube, the esophagus, whence it passes into the stomach. This is a rather large organ lined with chitin which is thickened and hardened in certain places called the *ossicles*

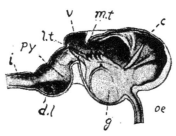

Fig. 74.—Stomach or "gastric mill" of the crayfish cut through the middle. *c*, cardiac regions of stomach; *d.l*, duct from the liver; *g*, gastrolith, or calcareous disk secreted by the walls of the stomach; *i*, intestine; *l.t*, lateral teeth of grinding apparatus; *m.t.*, median tooth; *oe*, esophagus; *py*, pyloric region; *v*, valve between cardiac and pyloric regions of stomach. (After Hatschek and Cori.)

which act as a sort of grinding apparatus. The posterior part of the stomach receives the ducts from two large digestive glands, commonly called the *liver*. These pour into the stomach a digestive fluid which acts upon the ground-up masses of food, making them capable of absorption into the blood. At the posterior part of the stomach there project into the cavity a number of hairs which act as a strainer, allowing only the finely divided food to pass backward into the intestine. The latter is a straight tube extending backward into the abdomen to open at the under side of the base of the telson.

Behind the stomach lies the large *heart* which is enclosed in a membranous sac called the *pericardium*. From the heart there arise a number of *arteries* which extend forward and backward and carry blood to all parts of the body. Blood enters the heart from the pericardium through three pairs of apertures, the *ostia*, which are provided with valves to prevent its return. The blood which has been forced by the beating of the heart, to all parts of the body, passes into spaces between the tissues, called *sinuses*, and finally collects in a large sinus lying along the ventral side of the body. From here it passes to the gills and then flows back into the pericardium. Thence it again passes through the ostia into the heart to repeat its journey. The blood of the crayfish is colorless and contains many small bodies called corpuscles, whose functions will be discussed in another chapter.

Much of the waste matter in the blood is got rid of by organs called from their color the *green glands*. They are situated on either side of the esophagus and open at the base of the large antennæ. The reproductive organs in mature individuals lie partly in the cephalothorax and partly in the abdomen. Their ducts in the male open at the base of the last pair of walking legs, in the female in the base of the second pair in front of the last. It is not difficult to distinguish the sexes of most of the higher crustaceans by the position of these openings.

The nervous system is quite similar in its fundamental features to that of the grasshopper. There is a *brain* over the esophagus giving nerves to the eyes and both pairs of antennæ, and cords passing ventrally on either side of the esophagus to a large *subesophageal ganglion* which supplies nerves to the mouth parts; from this ganglion a double chain of ganglia extends along the ventral side of the cephalothorax and abdomen.

Crayfishes are very widely distributed over the United States and occur also in many parts of the old world. There are many different species. Some of them inhabit ponds and streams; others are found in damp soil where they dig holes often to a depth of several feet. Some of the burrowing species heap up the dirt which is brought up in digging, so that it forms a tube or "chimney" over the hole. One species of crayfish which is found in Mammoth Cave, Kentucky, is blind; the eye-stalks remain, but the eyes have disappeared. As a rule crayfish are retiring in their habits and usually lodge under stones or in other dark and protected situations, although curiously enough they may be attracted by a strong light at night. Boys often catch them by letting down into the water a string with a piece of meat tied on one end. The crayfish siezes the meat and it does not occur to it to let go before being pulled out of the water.

The breeding season varies greatly in different species. The young after they are hatched have a strong instinct to cling with their chelæ to any object within reach, and for some time hang tenaciously on to the swimmerets of their mother. After a while the young leave their parent and shift for themselves. Like young grasshoppers, they shed their skin several times before reaching maturity. In the process of molting the skin splits between the thorax and abdomen and the crayfish slowly pulls its body and legs out of the old cases. Even the lining of the stomach and part of the intestine are shed also. Molting is a trying process for crustaceans in general and some die as the result of the ordeal. As the new skin is very soft the crayfish is not well able to protect itself and usually retires to some sheltered spot. Its usual pugnacity disappears as if it recognized its helpless plight. As the crayfish approaches its full size molting occurs much less frequently. Concerning the age reached by crayfish we know little except that they live at least several years. Crayfishes are used for food quite extensively in Europe and to a considerable extent in this country.

Among the nearest allies of the crayfish are the lobsters. The American lobster, *Homarus americanus*, which is closely allied to the European species, is found on the eastern coast of the United States as far south as Virginia. It may reach a length of two feet and a weight of twenty-five pounds. As it is a favorite article of food, it is caught in great numbers so that the larger individuals are now more rarely found. It is usually caught in a wooden cage called a lobster "pot," which is so constructed as to allow the lobsters to go in—which they are induced to do to obtain the bait—but which prevents their escape. Owing to the decreased yield of our lobster fisheries it has been made illegal to sell lobsters of less than a certain length. At various places on the coast, lobster hatcheries have been erected. Here the eggs are kept in jars of running water until they hatch, when the young are carried out to sea. How greatly the supply of lobsters has been increased by this method is a matter of great uncertainty. The young lobster makes its first appearance in a larval form very different from the adult and passes through a long metamorphosis before attaining its final form.

Somewhat more distant relatives of the crayfish are the various species of prawns and shrimps. These, like the lobsters, are caught in large quantities for the gratification of the human appetite. The hermit crabs are noteworthy for their common habit of living in the coiled shells of mollusks into which they can more or less completely withdraw. The anterior part of the body and the anterior appendages which are habitually exposed to the impacts of the outer world are hard, but the abdomen which remains protected within the deeper part of the shell has become soft and fleshy. The terminal appendages of the abdomen have been modified into organs to enable the crab to retain its hold of the shell. In front of these the appendages are generally absent on the right side of the abdomen, an indirect consequence, probably, of the spiral twist to which the abdomen is subjected. In the males the terminal appendages of the abdomen are often the only ones present. Hermits hang on to their shells with great tenacity and may even allow their bodies to be torn in two before releasing their hold. When the hermits outgrow their shells they hunt larger ones and a hermit often tries to take possession of a shell that is occupied by another individual.

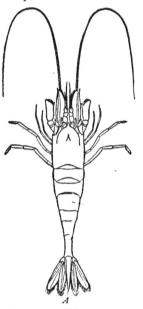

FIG. 75.—The common shrimp, *Crangon vulgaris.*

The true crabs usually have a short, broad carapace and a small abdomen which is folded under the cephalothorax where it fits neatly into a concave space. The crabs which are best known are those prized for food, such as the blue swimming crab of the Atlantic coast; but there are many species which have very interesting habits. Among these may be mentioned the fiddler crabs, the males of which have an enormously developed cheliped which is held horizontally across the front of the body. These active creatures live in holes dug in the sand or mud near the water's edge. They run with a good deal of agility and usually make for their holes at the appearance of danger, the males holding up their large claws in a threatening manner while beating a lively retreat.

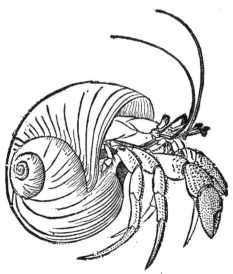

Fig. 76.—*Pagurus bernhardus*, hermit crab. (From Emerton.)

The spider crabs, on the other hand, are generally very sluggish in their movements. They usually have long,

Fig. 77.—A crab, *Panopeus chilensis*. (After M. J. Rathbun.)

slender legs and a body narrowed and pointed in front. In some species the back is covered with a growth of seaweed, sponges, hydroids, etc., so that it is difficult to detect the animal in its natural surroundings. It is a curious fact that these growths may be planted by the crabs themselves. A spider crab placed in an aquarium with bits of seaweed will snip off pieces with the pincers, reach back and place them among the short, hooked spines and hairs on its carapace where they become attached and grow. A crab may deck itself out with bits of paper in a similar manner. It has been found by experiment that this wonderful instinct does not depend upon the brain, as a crab will proceed to disguise itself in the same way after the brain is entirely destroyed.

The foregoing crustaceans, however much they differ in external appearance, agree in having many characteristics common by which they are grouped in one order, the Decapoda. The term has reference to the possession of ten legs which is a general feature of the group. The Decapoda also have a carapace, stalked eyes and gills on the cephalothorax.

FIG. 78.—A sow bug, *Porcellio*, enlarged. (After Essig.)

Another order standing somewhat lower in the scale is the Isopoda. In this group the typical number of legs is fourteen; there is no carapace, the eyes are not stalked but sessile, and the gills are formed by modifications of the appendages of the abdomen. There are many marine isopods, some of which are parasites of fishes, while some very degenerate forms prey upon other crustaceans. One species, *Limnoria lignorum*, is a serious nuisance to man, as it bores into the piles of wharves and so riddles them with its burrows that they soon become useless. A few isopods, Asellus and its allies, occur in fresh water, but the most familiar ones, our common sow bugs and pill bugs, live in damp situations upon land and have become adapted to breathing air. These terrestrial forms are often found under logs and stones, in damp cellars and around old buildings. They are mostly vegetarians, but do not disdain a little meat occasionally. For the most part, however, they are content to fill themselves up on such apparently unattractive pabulum as partially decayed wood. While they sometimes attack tender young plants they are for the most part harmless creatures and may even be of benefit in a small way as scavengers.

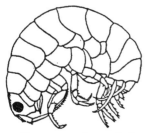

FIG. 79.—A sand flea, *Talorchestia*.

The nearest allies of the isopods are the amphipods which agree with the isopods in having sessile eyes, no carapace and the typical number of fourteen legs. The gills, however, are attached to the thorax. Most amphipods are marine, but there are many fresh water species, and a few terrestrial ones called sand fleas commonly found on sandy sea beaches.

At a first glance no one would classify the barnacles with the Crustacea and up to less than a century ago even zoologists classed them, along with clams and snails, among the Mollusca. This was doubtless done on account of the hard shell with which the body of most barnacles is surrounded. It was later found that barnacles hatch from the egg as a *nauplius*, a common larval form in other groups of crustacea. The nauplius is a free swimming larva with a median eye and three pairs of appendages. The barnacle nauplius as it grows undergoes a series of molts accompanied by considerable changes of form, and finally settles down and attaches itself by its head to some object and gradually assumes the form of a small barnacle.

The older naturalists classified animals mainly on the basis of external form instead of internal structure. Had they studied the organization of the part of the barnacle within the shell they would have found that the animal really resembles other crustacea even in its adult state. It has mouth parts consisting of mandibles and maxillæ; the feathered appendages which it continually thrusts out and withdraws into its shell while it is in the water are the thoracic legs richly supplied with hairs for entang-

FIG. 80.—A group of barnacles. (After Pilsbry.)

ling the small creatures used as food. Some barnacles, such as the common goose barnacle, are provided with a flexible stalk, while others such as the acorn barnacles have the shell attached directly to some other object. Barnacles are frequently attached to the hulls of ships where they may be so numerous as to greatly impede the vessel's movements. Some species attach themselves to the skin of whales. others to sea turtles. Some members of the barnacle group have become parasitic and have degenerated to such a degree that the adults would never be taken for crustaceans at all were it not for our knowledge of their life history. One of the most extreme cases of degeneration through parasitism that is known is furnished by Sacculina, a parasite on crabs. This parasite appears as a fleshy mass attached to the body of the crab commonly under the abdomen. The Sacculina sends rootlets into its host which penetrate and draw nourish-

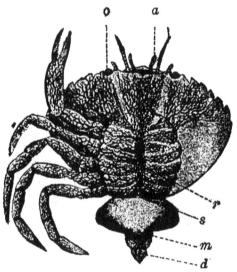

FIG. 81.—*Sacculina carcini* attached to *Carcinus mænas*, whose abdomen is extended. *m*, shell opening; *r*, network of roots ramifying the crab; *s*, stalk; *a, o, d*, antennula, eye and anus of the crab.

ment from nearly all parts of the body. The parasite has no appendages, segments, or any external structures characteristic of crustaceans, and the alimentary canal and most other organ systems, except the reproductive organs, have disappeared. The Sacculina has been reduced to an apparatus for absorbing food from its host and producing eggs and sperms. The eggs hatch into nauplius larvæ which pass through early stages of metamorphosis closely resembling those of typical barnacles.

Finally the larva becomes attached by its antennæ to a crab. Then it begins to lose its specialized organs, pushes its branches into the host, and becomes a fleshy and almost structureless mass. Knowledge of the early development of an animal sometimes affords a clue to its true relationship which could be ascertained in no other way.

There are multitudes of species of the lower crustacea which are usually of small size. Some of the most common are the water fleas (Daphnia and allied forms) which are frequently found in fresh water. The minute copepods which are abundant both in fresh water and in the sea form an important factor in the food supply of aquatic animals, especially fishes.

CHAPTER XI

THE MOLLUSCA

The Molluscs include such animals as clams, snails slugs, devil-fish and their allies. One of the most widely distributed and generally available of the molluscs is the fresh-water clam of which there are numerous species in the lakes, ponds and streams of North America. We shall therefore use it as an introductory type. The two valves of the shell by which the body of the clam is enclosed are secreted by a fold of the body wall called the *mantle* which hangs down on either side of the body. The shell grows in thickness by additions from the mantle to its inner surface, and in area by additions to the edge, the concentric lines visible on the outside of the shell indicating periods of growth. Where the valves are joined together there is a thick, elastic body called the *hinge ligament* which acts as a spring to open the shell. The shell is closed by two muscles called *adductors* which run from one valve to the other. The inner surface of the empty shell shows the marks made by the insertion of the two muscles near either end, and also the line of attachment of the mantle to the shell.

By removing one valve of the shell one may see the gills, two pairs of which hang in the mantle cavity, a pair on either side of the body. These gills are made of numerous fine filaments joined together so as to form broad plates or *lamellæ*, hence the term *lamellibranch* which is applied to the group to which the clam and other molluscs with bivalved shells belong. Each gill is composed of two lamellæ which separate above to form a canal opening outward between the folds of the mantle at the posterior end of the body. The surface of the gills is covered by fine cilia which beat so as to cause a current of water to flow in through small orifices between the filaments and into the canals above the bases of the gills, and thence out through

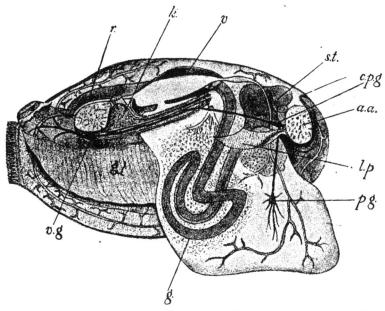

FIG. 82.—Structure of the clam Anodonta, *a.a*, anterior adductor muscle for closing the two valves of the shell; *c.p.g*, cerebropleural ganglion or brain; *g*, intestine coiled in the foot; *g.l.*, gill; *k*, kidney; *l.p.*, labial palp; *p.g.*, pedal ganglion; *r*, rectum; *s.t.* stomach; *v*, ventricle of heart giving rise to arteries at each end; *v.g.*, visceral ganglion just below the posterior adductor muscle. (After Rankin.)

the opening between the mantle lobes, called the *exhalent siphon*. Water flows into the mantle cavity through the *inhalent siphon* which is situated just below the exhalent one. Place some colored fluid in a dish of water containing a clam, and you may see it drawn into the one opening and expelled at the other.

The lower part of the body projects in front into a muscular organ, the *foot*, which can be protruded between the valves of the shell. The clam makes use of the foot in burrowing into the mud where it commonly lives with the siphons exposed to the water. The water, which in passing through the gills subserves the function of respiration, affords also the means of bringing the animal its food which consists of microscopic organisms and other fine materials swept in by the ciliary current. The solid bodies are carried by ciliary action into the mouth which is situated between two pairs of flaps called the *labial palpi* at the anterior end of the body. The mouth leads by a short tube to the *stomach* which receives the ducts from a large greenish digestive gland commonly called the *liver*. The stomach leads to the narrow *intestine* which after coiling about in the body opens near the posterior end of the body where its contents are carried out through the exhalent siphon.

The clam is furnished with a *heart* consisting of a median *ventricle* and two lateral *auricles* lying in a space called the *pericardium* in the dorsal side of the body. The intestine passes through this pericardium and is surrounded by the ventricle of the heart. The beating of the heart carries the blood through arteries to various parts of the body. On its return it goes through the gills where it takes up oxygen and loses a part of its waste products, and then passes into the auricles and thence into the ventricle. Just below the pericardium is a pair of dark-colored excretory organs or *kidneys* which open at one end into the pericardium and, at the other, to the outside of the body.

The nervous system of the clam consists of three main pairs of ganglia connected by nerve cords or *commissures*. The *cerebral* or *brain ganglia* lie over the mouth. These are connected by long commissures to a pair of large *visceral ganglia* just below the posterior adductor muscle.

There is another pair of commissures extending from the brain to a pair of *pedal ganglia* in the foot.

A considerable part of the compact body of the clam is made up of the sex organs which discharge their cells near the opening of the kidney. The eggs when set free fall into the spaces between the lamellæ of the gills where they undergo their early development; the young larvæ are then carried out through the excurrent siphon and live for some time at the bottom of some body of water. For the next stage of its development the young clam is dependent upon becoming attached to the gills or fins of some fish.

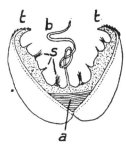

FIG. 83.—Larva of clam Anodonta. *a*, adductor muscle; *b*, byssus thread; *s*, sensory hairs; *t*, teeth for attachment to host.

When this opportunity presents itself the young clam closes the valves of its shell over the tissues of its host and hangs there. Later it becomes more or less completely overgrown by the surrounding tissues of the fish, much as a gall insect is enclosed in the tissues of a plant gall. Finally, the young clam breaks out of its enclosure, settles down in the mud, and begins the regular routine life of its parents. The shells of fresh-water clams are much used in the making of buttons. Occasionally they yield pearls of considerable value. A pearl is a calcareous secretion of the mantle which accumulates around some foreign body, commonly a parasitic

FIG. 84.—Pearl oyster from Ceylon, showing pearls on inner surface of shell.

worm. Pearls may be formed by various kinds of molluscs, and in some places pearl fisheries form an important industry.

Several species of clams found on the seashore are much used as food. One of these, *Mya arenaria*, the common long-neck clam, is obtained by digging in muddy beaches at low tide. The two joined siphons in this species constitute a long tube which projects upward as the clam lies buried in the mud. When the clam is disturbed it frequently reveals its presence by squirting water out of its siphon as it closes the valves of its shell.

FIG. 85.—A scallop shell, Pecten.

Mussels are generally found upon rocks to which they attach themselves by a series of threads called the *byssus* which is secreted by a gland in the foot. The common scallop, Pecten, has the somewhat unusual habit of swimming by alternately opening and closing the valves of the shell. The most important bivalves are unquestionably the oysters which are extensively cultivated in various parts of the world. In its early stages the oyster is a free-swimming larva; later it settles down and becomes attached by the left valve of its shell. Oysters are planted and cultivated in oyster beds. These beds are especially numerous in Chesapeake Bay which has provided over 25,000,000 bushels of oysters a year. A very aberrant bivalve and one which looks more like a worm than a mollusc, is the Teredo which has the habit of boring into the piles of wharves and bottoms of wooden vessels where it does a great deal of damage by riddling the wood with its holes.

One large division of the Mollusca, the Gastropoda, is

characterized by an asymmetrical and usually coiled body. Most of the species live in a coiled shell, but in some forms, such as the limpets, the shell is a sort of cap; in the slugs it may be reduced to a small rudiment imbedded in the mantle while in some forms no trace of the shell remains in the adult although a shell gland is present in the embryo. The Gastropoda are common in the sea, in fresh water, and on land. One can in most places easily obtain a fairly typical gastropod in the familiar garden snail, Helix. The body is furnished below with a broad, flat, muscular base, the foot, on which the animal creeps. The

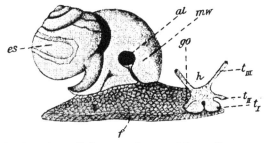

FIG. 86.—Body of snail in creeping position after removal of shell. f, foot; go. opening of sex organs; h, head; al, opening into lung cavity; mw, wall of mantle; es, visceral sac; t_I, t_{II}, and t_{III}, tentacles, the upper pair with eyes at the tip. (From Meisenheimer.)

head bears, in addition to the short feelers over the mouth, a pair of long, retractile tentacles at the end of which are the eyes. When the latter are irritated they may be drawn into the tentacles much as a person could pull in the end of a finger of a glove by a string attached to the inner side of the tip. In the mouth of the snail there is a ribbon-like structure armed with rows of minute chitinous teeth which are used in rasping off bits of food. On the right side of the body near the head are the openings of the sex organs, and further back a larger aperture which leads to the breathing cavity or lung. While in the marine gastropods this cavity contains gills which are used for respiration in the water, in the land snails and many fresh-water species the gills have been lost and the surface of the cavity adapted for breathing air. The garden snail feeds mostly on the tissues of plants, but it will also devour meat and various other kinds of food.

FIG. 87.—A pond snail, Physa, having a reversed spiral coil.

It travels most at night leaving evidences of its journeys in the form of slime tracks which result from the mucus secreted by the foot. During the winter, and sometimes in periods of drought, the snail draws into its shell and secretes a porous limy substance over the mouth of the shell called the *epiphragm*. Thus sealed up, the snail lives in a dormant state until the advent of more favorable conditions of life.

Many of the lung-breathing relatives of Helix live in fresh water. The common pond snails, Limnæa, Physa and Planorbis may easily be kept in aquaria where one may watch their many interesting peculiarities of behavior. These forms usually come to the surface for air, and, after filling the lung, descend. One curious habit of many pond snails is the spinning of mucus threads from the bottom to the surface film. The snails crawl up and down upon these threads in their periodic excursions to the surface for air.

FIG. 88.—*Conus eburneus.*

The sea abounds in gastropods of the most varied forms, sizes and habits. Some of these are carnivorous and prey upon other molluscs. One often finds a bivalve with a round, smooth hole bored through its shell. This tells the story of some carnivorous gastropod which had bored into the helpless bivalve with its rasp and devoured its soft parts. One species called the "oyster drill" destroys large numbers of oysters in this way.

The chitons which are allied to the gastropods have a broad, creeping foot and a dorsal shell composed of a row of eight pieces. They are often found on rocks at low tide. The Cephalopoda differ greatly in appearance from the other mollusca. Their name is derived from the fact that the foot is produced into a number of arms (8 or 10) surrounding the head. In most cephalopods these arms are furnished with rows of suckers which are used for retaining hold of objects. The cephalopods were represented in former ages of the earth's history by vast numbers of varied forms. Some of these, the ammonites, had a coiled, chambered shell which was often beautifully sculptured. The orthoceratites had a straight, chambered shell which in some species reached a length of fifteen feet. Of the forms now living within a shell, the chambered nautilus is the only

FIG. 89.—*Terebra babylonia.*

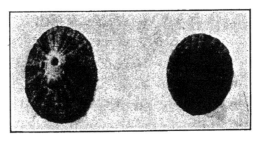

FIG. 90.—Limpets, a key hole limpet at the left.

survivor. It is found in deep waters in the Pacific and Indian Oceans, but its much prized shell is sometimes cast up on the shore. In the so-called paper nautilus there is a thin shell which is formed, not by the mantle, as in the chambered nautilus and other molluscs, but by the expanded posterior arms; it occurs only in the female where it serves as an egg case.

Most of the other living cephalopods have a relatively small shell which is overgrown by the mantle. This may be calcareous as in the cuttle fish which furnishes the "cuttle fish bone" that we often give to canaries, or it may be chitinous as in the "pen" which lies along the dorsal side of the body of the squid. It is commonly said that the squid carries its own pen and ink; the animal is furnished with an ink-sac containing a black fluid which may be discharged in times of danger, producing a black cloud in the water which facilitates the escape of the animal. There are many species of squid living in various parts of the world. One of the forms most easily obtainable for study is the small *Loligo pealii* found along our Atlantic coast. Of the ten sucker-bearing arms extending in front of the head two are longer than the others. In the center of the circle of arms is the mouth with its two strong, horny jaws resembling the beak of a parrot. As in other cephalopods the head bears two large, well-developed eyes. The squid swims backward by suddenly expelling water from its mantle chamber through the siphon which is a short tube below the head. This siphon can be turned in various directions so that the reaction of the expelled water may cause the animal to turn in different ways. The mantle cavity contains the gills of which there are two pairs. Squids are very active animals, living on small creatures which they capture by means of their arms. They are remarkable for the rapid changes of color which may pass over the skin, especially when they are disturbed. There are a few species called the giant squid which attain a very large size, with a body nine to ten feet long and with the longest pair of arms reaching a length of forty feet.

Stories of these monsters attacking ships are quite without foundation.

Unlike the squids, which are very graceful and attractive animals, the devil fish, or octopi, are apt to inspire one with a feeling of repulsion. Nevertheless they are very interesting creatures and the feeling of repulsion, as in so many other cases, will tend to disappear upon closer acquaintance. As the name Octopus implies there are but

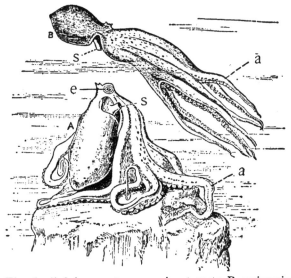

Fig. 91.—The devil fish or octopus. *A*, at rest; *B*, swimming; *a*, arms with suckers; *e*, eye; *S*, siphon. (From Cooke, after Merculiano.)

eight feet or arms; these are relatively long and very strong and they enable the octopus to overcome quite powerful adversaries. The larger species whose arms may reach a length of fourteen feet might be dangerous to man who is a comparative helpless creature when in the water. There are many stories of human beings being attacked and overcome by devil fish, but they are mostly due to the proverbial mendacity of fishermen. Like the squid the devil fish are sometimes used as food, but they have never won their way to general favor.

CHAPTER XII

THE ECHINODERMS

In the Echinoderms, or spiny-skinned animals, Nature has worked out a peculiar type of organization very different from what is found anywhere else in the animal kingdom. One conspicuous feature of most echinoderms is their apparently radial symmetry. The parts of the body are arranged about a central axis instead of merely on two sides of a median plane as in insects and vertebrates. This radial structure led the older naturalists to class the echinoderms along with the jelly-fish and their allies in a group called Radiates, but it is now known that jelly-fish and echinoderms are but very distantly related. Even the radial structure on which the alliance was based is now known to be secondary in the echinoderms and derived from a primitive bilateral symmetry which can still be traced in the position of certain organs of the body. Without exception the echinoderms are confined to the sea.

One large division of the Echinoderms consists of the Asteroids, or starfishes, of which there are many species of various forms and colors. Very commonly there are five rays which extend from a central disk, but in some cases there may be six, and in a few, even over twenty. The body wall is hardened by a deposit of lime, but not to such a degree as to prevent more or less movement of the rays. The principal organs of locomotion are the *tube feet* which project in rows through the under sides of the rays where they are lodged in grooves. In most species the tube feet end in adhesive suckers. By extending, attaching, and then contracting the tube feet the starfish manages to pull itself along over the bottom, or even to climb up vertical surfaces. The tube feet also serve to seize prey and carry it toward the mouth.

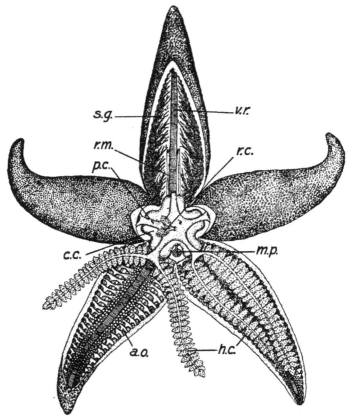

FIG. 92.—Anatomy of Starfish, *Asterias vulgaris*. *a.o*, skeletal pieces of arms; *c.c*, cardiac cœca or pouches of stomach; *p.c.*, pyloric cœca; *h.c.*, hepatic or "liver" cœca; *r.c*, rectal cœca; *r.m*, muscles; *s.g*, reproductive glands; *m.p.*, madreporic plate through which water enters into the water vascular system. (After Coe.)

Although the starfish is a very harmless and innocent looking creature, it is able to overcome and devour quite large animals. When an object is brought near the mouth the stomach may become actually thrust out of the body and wrapped around the object which it slowly digests.

Oysters, clams, and other mollusks which are too large to be taken into the body are often digested in this way, after which the stomach retracts leaving the empty shells. Starfishes often do much damage to oyster beds, and they are very difficult to get rid of.

Scattered over the body of many kinds of starfish are numerous minute bodies called *pedicellariæ* which look like miniature pairs of forceps. These organs have usually two jaws which open and close by means of special muscles. They frequently catch hold of objects coming in contact with the starfish, and are thus serviceable in capturing prey. A live fish, longer than the diameter of the starfish, has been observed to be held by these minute organs until it was conveyed by means of the tube feet within reach of the extensile stomach.

FIG. 93.—An ophiuran, or brittle star. A, disk seen from below showing the mouth.

The power of regeneration is very well developed in most starfishes. They will easily restore missing rays and even considerable parts of the disk, and there are a few species in which a single ray may give rise to an entire individual.

The Ophiurans, or brittle stars, may be distinguished from the starfishes by their circular disk which is clearly set off from the slender arms. When an arm is seized it is frequently cast off by a violent muscular contraction, hence the term brittle star.

In the Echinoids or sea urchins the body is generally circular or oval in outline and covered with movable spines. Generally also sea urchins are provided with tube feet having adhesive terminal disks which are used much as they are in the starfish. The spines are joined by a

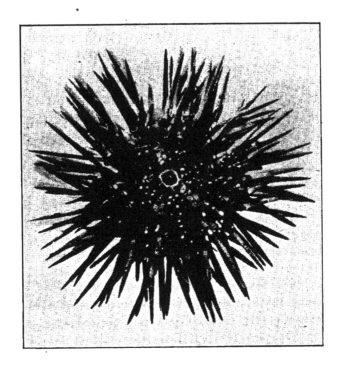

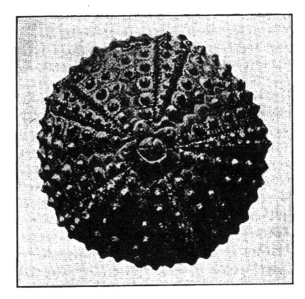

Fig. 94.—A sea urchin, Arbacia, and its shell with the spines removed. (After Clark.)

sort of ball-and-socket joint to rounded prominences on the shell, and they can be moved in different directions by a short ring of muscle fibers attached near the base.

FIG. 95.—"Aristotle's lantern," the chewing apparatus of the sea urchin.

Both spines and tube feet may be employed in locomotion, but the spines also serve as organs of protection. After removing the spines the shell of the sea urchin may be seen to consist of several regularly arranged series of plates. The plates are perforated where the tube feet are attached. The sand dollars are greatly flattened sea urchins with very short spines.

The sea cucumbers, or Holothurians, have an elongated body with a flexible and usually somewhat leathery wall. They are generally somewhat flattened on the side upon which they crawl, and the mouth is surrounded by tentacles which are used in the capture of food. With rare exceptions, there are tube feet projecting through the body, and these are usually arranged in five double rows. Some of the holothurians are extensively used by the Chinese as food.

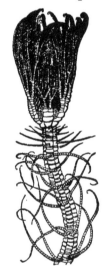

The Crinoids, or sea lilies, are usually attached by a jointed stem, although there are a few species that swim freely through the water. The usually cup-shaped body bears several branching arms which are furnished on the upper side with grooves which lead to the mouth. Very numerous

FIG. 96.—A crinoid or sea lily.

species existed in past ages of the earth's history. Their remains are common in the older rocks, but there are a comparatively few forms living to-day.

The eggs of echinoderms are generally shed directly

in most cases into free-swimming, ciliated larvæ, very different from the mature forms. There are different types of larvæ characteristic of the different groups of echinoderms. In the Ophiurans and sea-urchins the common larval form is furnished with long, ciliated arms and is known as a *pluteus*. After a short, free life the larvæ settle down and undergo a complicated metamorphosis in changing into the mature form.

CHAPTER XIII

THE RINGED WORMS OR ANNELIDS

The term worm is one of wide and somewhat indefinite significance. The old group called "Vermes," which is Latin for worms, constitutes what Professor Haeckel has called the great lumber room of Zoology, for it includes animals of the most diverse kinds, with little in common except that they do not belong to other groups. Nowadays zoologists parcel the Vermes out into a number of different phyla. One of the largest of these phyla is the Annelida. These are worms having the body divided into more or less similar annuli or segments, and provided generally with a body cavity or space between the digestive tube and the body wall.

There are a great many marine species, some of which are free, active, carnivorous creatures; others are sedentary, living in tubes and generally subsisting on small organisms. There are many annelids which inhabit fresh water or burrow in the soil. The latter are commonly known as earthworms or angle worms. There are a great many species of earthworms in various parts of the globe, one of the most common and widely distributed being *Lumbricus terrestris* which is frequently found in gardens and fields both in Europe and in North America. In this species, which we may take as a type, the body is composed of a remarkably uniform series of segments. Just over the mouth there is an incomplete segment called the *prostomium*. At about the anterior third of the body a few of the segments of the mature worm are somewhat thickened, forming the *clitellum*, which has an important function in relation to reproduction. Nearly all the segments are furnished with minute bristles, or *setæ*, which are arranged in four double rows. These may be seen with a hand lens, or felt with the finger, especially when it is rubbed from behind forward over the ventral surface. These setæ can be protruded or withdrawn into

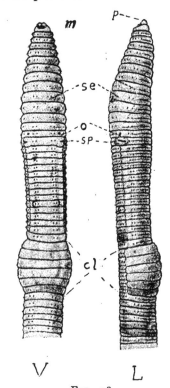

FIG. 97.   FIG. 98.

FIG. 97.—Earthworm from dorsal side. *cl*, clitellum; *p*, prostomium. (After Hatschek and Cori.)

FIG. 98.—*V*, ventral; *L*, lateral view of earthworm. *cl*, clitellum; *o*, opening of oviduct; *sp*, opening of sperm duct; *m*, mouth; *p*, prostromium. (After Hatschek and Cori.)

special sacs, and they materially assist the worm in locomotion.

The body of the worm is divided by a number of *septa*, or partitions, extending from the digestive tube to the body wall, each septum forming the boundary between two segments of the body. The digestive tube extends straight from the mouth to the anus in the last segment, and is divided into a number of parts. Behind the mouth

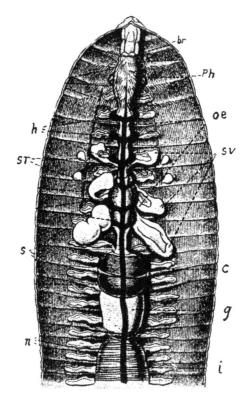

Fig. 99.—Internal organs of earthworm. *br*, brain; *c*, crop; *g*, gizzard; *h*, hearts arising from the dorsal blood vessel; *n*, nephridia; *oe*, esophagus; *ph*, pharynx; *sr*, seminal receptacle; *sv*, seminal vesicles. (After Hatschek and Cori.)

is a muscular *pharynx* which is used in drawing things in. This is followed by a short *esophagus*, leading to a thin-walled enlargement, called the *crop*, closely behind which is a thick-walled muscular organ, the *gizzard*, which serves as a grinding apparatus. Behind the gizzard is the *in-*

testine which extends with little modification to the last segment. Attached to the sides of the esophagus, and really consisting of outpocketings of this organ, are the *calciferous glands* which secrete limy crystalline bodies supposed to neutralize the acids contained in the food. Earthworms swallow dead leaves and other organic substances along with large quantities of soil, and digest whatever food there may be contained in this material. Rich soil with a considerable proportion of vegetable matter is therefore a favorite haunt for these animals. Absorption probably takes place mainly in the intestine.

The earthworm has a well-developed circulatory system containing red blood. The principal parts of this system consist of a *dorsal vessel* extending the length of the body above the digestive tube, a ventral blood vessel running below the intestine, and, lateral vessels extending from these to adjacent parts. The dorsal and ventral blood vessels are connected in front of the crop by five pairs of segmentally arranged vessels which surround the esophagus. These are called "hearts" because their pulsations help to propel the blood. The dorsal vessel contracts from behind forward, forcing most of the blood through the hearts into the ventral vessel where it is carried posteriorly.

The earthworm has very odd organs of excretion called *nephridia;* these are more or less coiled tubes, of which there occurs a single pair in most of the segments of the body. At its inner end, the nephridium is furnished with a ciliated funnel which passes through the anterior septum of its segment to open into the cavity in front. The outer end of the tube opens by a small pore through the side of the body. Waste material swept into the ciliated funnel or secreted by the walls of the tube is carried to the outside.

The earthworm's nervous system consists of a small brain over the pharynx, connected by commissures with a double chain of ganglia extending along the ventral side of the body, there being in each segment a single pair of ganglia which supply the adjacent parts with nerves. While earthworms have no well-defined eyes, they are very sensitive to light and tend to keep in dark situations.

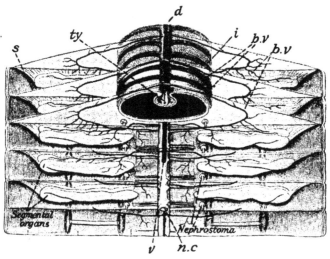

Fig. 100.—A few segments of the earthworm with the dorsal side cut away and showing the intestine cut through. *bv*, blood vessels; *d*, dorsal blood vessel; *i*, intestine; *n.c.*, nerve cord; *s*, septa; *ty*, typhlosole or fold projecting into the dorsal side of the intestine; *v*, ventral blood vessel. (After Hatschek and Cori.)

They will crawl away from the light; and when light is flashed on them at night when they are partly outside of their burrows they very quickly withdraw. Earthworms are very sensitive to chemical and mechanical stimulation. A slight jar may cause them to retreat quickly into their burrows.

The reproductive system of earthworms is very complex. Both male and female organs are located in the same individual but the eggs are nevertheless generally fertilized by sperm derived from another worm. The eggs are laid in a cocoon secreted by the clitellum. When the cocoon is formed it is slipped forward over the head, receiving the eggs as it passes the mouths of the oviducts, and deposited usually in a damp place.

Earthworms retreat from places that are very warm or dry; they are much more apt to come to the surface in damp or rainy weather when they may leave their burrows. At such times we may find them strewn about upon sidewalks and various other places much to the delight of robins and many other birds that prey upon them. Earthworms may be found even on the roofs of houses where many people suppose that they must have rained down. As a matter of fact they crawl up the sides of the house, as you may readily see them do if you give them the chance.

Earthworms have remarkable powers of regeneration. If a considerable part of either end is removed the worm will after a time regenerate the missing segments. Some of the aquatic relatives of the earthworm regularly multiply by fission in addition to developing from fertilized eggs.

As Darwin has shown in his interesting book on "The Formation of Vegetable Mould through the Action of Worms," earthworms play a very important part in the production of fertile soil. They burrow to a depth of several feet and bring to the surface a large amount of material that has passed through their bodies in their "castings" which accumulate around the mouths of their holes. These castings may be seen in any region in which the earthworms have recently been burrowing. They are washed away by rains or blown by winds; and, as Darwin has estimated, their removal may produce considerable change in the surface of the soil. As a consequence of bringing up soil from below the surface, rocks and other objects tend slowly to settle down and eventually become buried. Darwin observed a stony field which had thus become changed "so that after thirty years a horse could gallop over the compact turf from one end of the field to the other, and not strike a single stone with its shoes." Monuments and old buildings tend to settle slowly where they are undermined by earthworms, and in time may be-

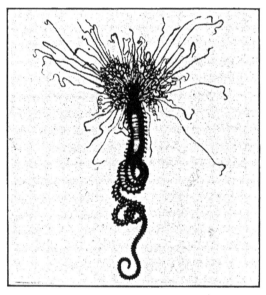

Fig. 101.—A tube-dwelling marine annelid. Note the branched gills at the anterior end and the thread-like cirri by means of which the worm entangles the small organisms that provide its food. (After Quatrefages.)

come completely buried. In an average field Darwin calculated that the amount of dirt carried to the surface by worms in one year would form a uniform layer one-fifth of an inch in thickness. Earthworms are thus continually plowing the ground, and although their operations may seem slow they may effect great changes in the course of centuries.

Related to the earthworms, although having very different habits of life, is the group of annelids called leeches.

The body of a leech is generally flattened and provided with a sucker at each end by means of which it adheres to various objects. Leeches crawl by a looping motion. Some species are furnished with teeth, especially those which live by sucking the blood of higher animals. Of the "blood suckers" the common medicinal leech is best known, since it was long used for bleeding patients, and was extensively raised in swamps and ponds especially prepared for leech culture. After a full meal of blood, a medicinal leech may live several months without food. Some species of leech attack small animals such as worms and snails; others, and especially the few that live in the sea, live upon the bodies of fish, and in the tropics there are land leeches which are troublesome pests to animals and man.

CHAPTER XIV

THE ROUND WORMS AND FLAT WORMS

The round worms, or nematodes, have an unsegmented and nearly cylindrical body commonly tapering toward one or both ends. Many species live in the soil or in decaying organic matter, while numerous others are parasitic in the bodies of animals. Some of the species, such as the large round worm of the horse, *Ascaris megalocephala* and the related species, *Ascaris lumbricoides* found in pigs and sometimes in man, attain a length of several inches. Others are of almost microscopic size, such as the vinegar-eels which are very frequently seen in cider vinegar. These forms are easily obtained and when observed with a microscope the principal internal organs may be seen in their semi-transparent bodies. They are entirely harmless, and there is no need to be fastidious about taking them in with our food.

One of the most dangerous of the many nematode parasites of man is the Trichina (*Trichinella spiralis*). The worms are commonly taken into the body by eating insufficiently cooked pork, for the Trichina is a common parasite of the pig. In the pork the worms are in an encysted state in the muscle; when this is digested, the worms are liberated, after which they grow to maturity in the intestine where they produce new worms. These young worms bore through the intestinal walls and get into the blood vessels where they are carried to various parts of the body; they then work into the tissues, commonly the muscles, and there encyst. It is during the invasion of the blood vessels and tissues that the worms produce their greatest injury, and a great many deaths have been caused by them. Infected pork may contain as many as 80,000 encysted worms in a single ounce. As these *Trichinæ* may produce many more young in the human intestine a person may be infected with millions of these minute worms after eating raw pork. The Trichinæ being readily killed by heat, it is easy to avoid these parasites by not eating pork that is insufficiently cooked. One should be especially cautious about eating raw salt pork, or raw smoked ham (both of which are eaten by many people) because it has been shown that the Trichinæ are not killed either by the salt brine or by the process of smoking. As in so many parasites the Trichina requires two hosts, the eater and the eaten, in order to complete its life history.

FIG. 102.—Encysted Trichinæ. (From Leuckart.)

Another serious human parasite is the hookworm of the southern states. The young of this form live in damp earth and gain access to man by boring in through the skin. People who went with bare feet in infected districts often contracted what was known as "ground itch" which is now known to be caused by the young hookworm. When through the skin the worms are carried by the blood throughout the body and many get into the alimentary canal; here they grow to maturity and produce eggs which are passed out of the body, where they hatch into young worms. The latter live in the soil where they await an opportunity to get into their host. When in the human intestine the worms produce considerable disturbance to general health, but they may be expelled by giving the patient liberal doses of thymol which kills the worms, fortunately without greatly injuring the afflicted person.

Besides the numerous nematodes living within animals, there are several serious parasites of plants. From our human point of view the nematodes in general are a bad lot, for there is scarcely any species for which a useful function has ever been discovered. One form which is sometimes classed with the nematodes may be mentioned here, viz., Gordius, or the so-called "horse-hair snake." This worm which is popularly but wrongly supposed to come from horse-hairs that have fallen into the water is parasitic during the earlier stages of its life, generally in the body of some insect, and only becomes free in the mature state.

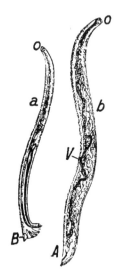

FIG. 103.—The hookworm. *a*, male; *b*, female; *o*, mouth; *V*, opening for discharge of eggs. (After Leuckart.)

The flat worms comprise a large phylum which is divided into three groups, the Turbellaria, the Trematodes or flukes, and the Cestodes or tapeworms. The most primitive group, the Turbellaria, are usually free-living animals, such as the planarians, which are commonly found under rocks in lakes and streams. The Trematodes are all parasitic forms characterized by having a forked intestine, and usually one or more suckers for attachment to the host. They are found in most vertebrate animals from fishes to man. One of the most injurious species is the common liver fluke *Fasciola hepatica* which frequently infects sheep and sometimes occurs in human beings. The flukes may reach a length of over an inch, and when they are present in considerable numbers in the liver, which is their usual abode, the host, whether a sheep or a man, has a very uncomfortable time. The life history of the liver fluke requires two hosts in addition to a period of life in the water. The eggs are passed out of the intestine and if they gain access to water they hatch into ciliated, free-swimming embryos. Should an embryo, in the course of its wanderings, encounter a water snail

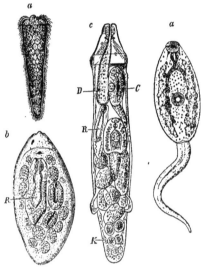

FIG. 104.—Developmental stages of the liver fluke *Fasciola hepatica*. *a*, free-swimming larva which develops in the body of a snail into a sporocyst *b*. The latter produces internally other larval forms, the rediæ *R*. *c*, a redia which contains still other rediæ *R* and a cercaria, *C* or final larval form. *d*, a cercaria. The cercariæ escape from the snail and swim freely in the water. (After Leuckart.)

it enters the body and there undergoes further development. The larval form so produced may give rise to numerous others by a process of parthenogenetic reproduction. Finally these larvæ leave the snail, swim about in the water, and frequently attach themselves to grass or weeds near the water's edge. Here they encyst. If now a sheep comes along, eats the grass that harbors the encysted larva, the latter develops into the mature form in the intestine, or liver of its host. Of course the chances are very small that any one embryo will make all the connections necessary for a successful life history. This circumstance is offset in part by the enormous number of eggs produced in the beginning, and in part by the fact that each embryo may produce many others provided it makes the first necessary connection with the body of a snail. Truly this seems to be a very roundabout and wasteful method of perpetuating the species, and calls to mind what the poet Tennyson said of Nature:

> "So careful of the type she seems,
> So careless of the single life."

No matter how many eggs or larvæ fall by the wayside so long as the race of liver flukes makes the journey from sheep back to sheep again. Each species of animal solves the problem of getting through the world in its own way; and the ways that are followed are often very devious and peculiar.

The Cestodes, or tape-worms, are also parasitic, and they have been addicted to such a life for so long that they have lost all traces of an alimentary canal. Instead of digesting their own food they live by absorbing the digested food in the alimentary canal of their host. With one possible exception, the Cestodes are parasitic in the adult state in the intestine of the vertebrate animals.

Most Cestodes are divided into a number of segments, or proglottids. In the larger human tape-worms which may reach a length of thirty feet there may be over 1000 of these segments. One end of the body is usually furnished with suckers and sometimes also with hooks for attachment to the wall of the intestine. Behind the attached end new proglottids are continually formed and they gradually increase in size and become more mature as they pass backward. Finally the proglottids become constricted off and pass out of the body. The mature proglottids contain fertilized eggs which may be set free either before or after the proglottid is expelled.

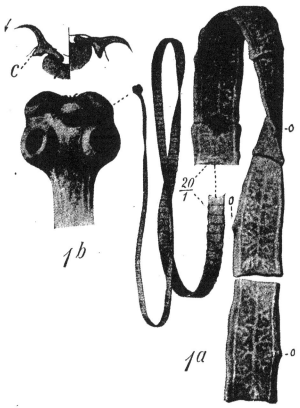

FIG. 105.—A human tape-worm *Tænia solium*. 1*a*, animal with a few segments enlarged at the right showing sex opening on one edge, *o*; 1*b*, attached end with suckers and circles of hooklets, the latter enlarged at *C*. (After a Pfurtscheller chart.)

The eggs do not develop directly into new worms; typically the eggs or embryos are taken into the body of some other animal; here the embryo bores its way through the walls of the intestine and becomes encysted in some part of the body, forming what is called a bladder worm (cysticercus or cysticercoid). In this state the worm develops only to a certain stage and does not attain maturity until it is taken into the body of some other animal. The bladder worm of one kind of human tape-worm, *Tænia solium*, occurs in the pig where the cysts frequently attain the size of a pea and may become considerably larger. The meat infested with these cysts is called "measly pork," and if such meat is eaten raw by man, the cysts will develop in his intestine into mature tape-worms. The bladder worm of another large species infesting man

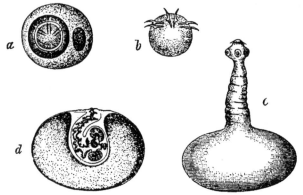

Fig. 106.—Development of the bladder worm of the tape-worm *Tænia saginata*. *a*, embryo within egg shell; *b*, free embryo; *c*, bladder worm; *d*, same with introverted scolex. (After Leuckart.)

occurs in cattle and the mature worm is acquired by eating insufficiently cooked beef.

Human beings may carry the bladder worms in their tissues and the mature worms in their intestine. If a person swallows the eggs of a tape-worm he may get bladder worms in his flesh. These may produce serious injury, especially if they lodge in the brain or some other delicate organ. One case is recorded of a bladder worm lodged in a woman's eye where its growth could be watched for several years.

While the bladder-worm stage of Cestodes is usually small there are a few species in which it is quite large. In

Coenurus cerebralis, which is found in the bladder-worm stage in sheep where it is often lodged in the brain, the cysticercus may reach the size of a hen's egg. It is a frequent cause of death to sheep.

Echinococcus in the bladder-worm stage may attain an enormous development. Some specimens reach the size of a child's head and weigh several pounds. Each cyst may produce a large number of worms. The tape-worm stage of this form is, curiously enough, a small worm of only four segments, usually found in the dog. The bladder-worm most commonly occurs in sheep, but there are many cases where it has been found in man, and has not infrequently been the cause of death. Echinococcus is common in Iceland and some parts of Australia where sheep and dogs are kept. Living in too intimate relations with dogs greatly increases the chance of becoming infected with this terrible parasite.

CHAPTER XV

THE CŒLENTERATES AND SPONGES

The hydroids, jelly fish, sea anemones, coral polyps, and their relatives constitute the phylum Cœlenterata. Primarily the Cœlenterates are radiate animals with their organs symmetrically disposed about a central axis, but there are some of the higher members of the group which have become to a greater or less extent two-sided or bilateral.

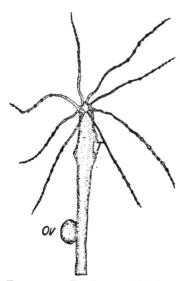

FIG. 107.—Fresh water Hydra. *Ov*, ovum; *T*, testes.

One can secure in almost any part of the country a living representative of this group in the common fresh-water Hydra which is frequently found attached to aquatic plants in ponds and streams. The tubular body of Hydra is furnished at one end with a variable number of tentacles (6–8 commonly) surrounding the mouth. At the opposite end is the foot which may become attached to objects by means of an adhesive secretion. The attachment is not permanent, however, as the Hydra can break loose at any time and crawl by a looping motion to another locality.

The internal structure of Hydra is very simple. The body may be regarded as a sort of sac composed of two layers of cells separated by a thin membrane. The inner layer is known as the *entoderm*, the outer as the *ectoderm*. The entoderm lines the large digestive cavity which extends the length of the body, and is continued also into the tubular cavities of the tentacles. Some of the entoderm cells are furnished with lash-like organs, or flagella, whose movements serve to circulate the contained material. The digestive juices poured out by the entoderm cells act on the food in the central cavity, but small particles of food may be engulfed within the cells themselves and digested there. Digestion is, therefore, extracellular, as it is in higher animals, and at the same time intracellular, as it is in animals still lower in the scale of life. The undigested residue of the food is ejected through the mouth.

The ectoderm cells in addition to forming an outer covering for the body are modified into gland cells, muscle cells, nerve cells, sex cells and nettling cells. The muscle cells endow the Hydra with its extraordinary contractility.

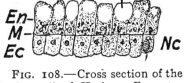

Fig. 108.—Cross section of the body wall of Hydra. *Ec*, ectoderm; *En*, entoderm; *Fl*, flagella; *M*, mesoglœa; *Nc*, nettling cell; *V*, vacuole.

At times the animal may be greatly elongated with its tentacles extended into fine threads several inches in length. Touch the animal one or more times; the tentacles will be reduced to mere stubs and the body contracted almost into a ball. There is no central nervous system such as occurs in higher animals, but a scattered network of nerve cells whose fine branches or nerves connect with various other cells of the body.

The most interesting cells of Hydra are the nettling cells which contain oval bodies called *nematocysts*. The latter consist of a hollow capsule containing a long thread spirally wound up on the inside. In response to certain stimuli the thread may burst out of its capsule and become extended with a good deal of force so that it may penetrate the tissues of animals, even those which are covered with a layer of chitin. The nematocysts contain a poisonous, irritating fluid which serves to paralyze the small animals that the Hydra feeds upon. Nettling cells are commonly furnished with a pointed projection, "the trigger," which is supposed when irritated to set off the discharge of the nematocysts. Nematocysts are especially abundant near the tips of the tentacles where they are most apt to come into contact with prey. The tentacles are also adhesive and small animals which come in contact with them are caught and drawn toward the mouth. The swallowing capacity of Hydra is enormous; animals considerably larger than the Hydra itself are successfully taken into the digestive cavity.

Reproduction in Hydra is effected in two ways, (1) asexually, by the formation of buds, and (2) sexually, through the production of eggs and sperms. In budding, an outpushing of both layers of the body wall occurs, tentacles are pushed out and a mouth breaks through at the outer end of the bud. Finally the bud constricts off at the base and forms a new free Hydra. Often many buds are found on one individual.

Both male and female sex cells are produced in the same individual. The eggs are produced, one or two at a time, on the basal part of the body. The sperms appear in a number of conical prominences nearer the oral end. The sperms are set free in the water and fertilize the large egg cell while it is still in the ectoderm of the body wall. In this situation also the egg undergoes its early development, but at a certain stage there is formed around it a chitinous and often spiny shell, or capsule, which serves to protect the egg after it is discharged, when it comes to lie at the bottom of a pond or stream. Enclosed in the shell, the egg can withstand periods of drought and other unfavorable conditions after which it may break out of the shell and complete its development.

Hydra has long been a classical object for the study of regeneration. It may be cut in numerous pieces each of which may form a new Hydra, and pieces from different individuals may be grafted together, much as nurserymen graft together different trees or shrubs. One species of Hydra, *Hydra viridis*, is remarkable for its green color.

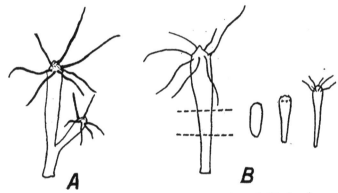

FIG. 109.—*A*, budding in Hydra; *B*, regeneration of Hydra from a small piece from near the middle of the body.

The green is due to the presence of small, unicellular green plants (algæ) in the cells of the entoderm. The plants ordinarily are not digested, but live on material derived from the host. As plants assimilate carbon dioxide which is a waste product of animals and give off oxygen which the animal uses in respiration, the association between the Hydra and the algæ is supposed to be to their mutual advantage. Thus we have an illustration of *symbiosis* which was briefly considered in a previous chapter.

There are numerous marine relatives of Hydra which are commonly called *hydroids*. Many of these are much branched and form colonies. In some of these there has come to be a division of labor between different individuals of the colony, some being specialized for catching prey (feeding hydroids); others, richly furnished with nettling cells, are set apart for protection (defensive hydroids); while others (the reproductive hydroids) are devoted entirely to reproduction. Very commonly the marine hydroids give rise by budding to a free-swimming generation of *jelly fish*, or *medusæ*. The larger part of a typical medusa consists of a *disk*, or *umbrella*, furnished with *tenta-*

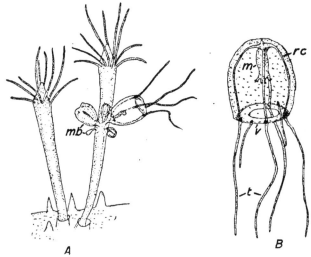

FIG. 110.—Hydroids *A* and medusa *B; m*, manubrium; *mb*, medusa buds; *rc*, radial canals; *t*, tentacles; *v*, velum. (After Allmann.)

cles along the outer margin. Hanging down from the middle of the lower side is the *manubrium*, at one end of which is the mouth. The latter leads to the stomach from which a number of canals (commonly four) radiate outward where they open into a circular canal near the margin of the umbrella. Jelly fish are usually transparent animals, frequently of very delicate and beautiful structure; they swim through the sea by contractions of the umbrella and live upon animals which they catch by means of their tentacles. Jelly fish produce sex cells which are discharged from the body and commonly develop, not into jelly fish, but into hydroids. We thus have what is called *alternation of generations*, jelly fish producing a hydroid (the asexual generation) which gives rise again to a jelly fish. Not all species pass through this alteration of generations. Just as there are hydroids which, like the fresh-water Hydra, have no medusa stage but develop directly from eggs produced by other hydroids, so there are medusæ which produce eggs that develop directly into medusæ and have no hydroid stage.

In the group of Siphonophores the division of labor which has been noted among certain hydroids reaches an extreme development. The siphonophores are all free-swimming or floating colonies made up of numerous individuals modified in various ways, some for swimming, some for taking food, others for protection, and still others for reproduction.

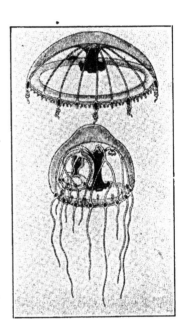

FIG. 111.—Two species of jelly fish from the Tortugas. (After Mayer.)

One of the largest species is the Portugese man-of-war which is furnished with a long oblong float filled with gas, by which the colony rides upon the surface of the waves. From the lower side of the float hang down clusters of variously modified individuals and very contractile tentacles which may extend to a length of fifty feet. These are richly furnished with nettling cells, and the poison they contain is so virulent that even a slight contact with a tentacle may produce considerable irritation.

The anemones are mostly sessile animals attached at the base or foot to rocks or seaweed. They are common on rocky parts of the coast and some forms may be exposed at low tide. Many species are remarkable for their beauty of form and coloring. The free end, or disk, is furnished with tentacles which are employed in catching prey and conveying it to the mouth. Anemones are more highly organized than the hydroids. The digestive cavity is divided by a number of partitions, the *mesenteries*, which extend from the body wall toward the center. In some species the edges of these mesenteries bear long extensile filaments, armed with nematocysts, which are capable of being thrust out of the body when the animal is irritated.

Closely related to the anemones are the corals. In a typical coral the body of the individual animal, or polyp, has the property of secreting about its base, a hard deposit of carbonate of lime, forming the so-called coral rock. As coral polyps commonly multiply by budding they may form an extensive society of individuals more or less closely associated with one another. The masses of coral rock which the polyps form may be increased almost indefinitely by the multiplication of the polyps and the accession

FIG. 112.—Portugese man-of-war. (After Agassiz.)

of new ones. Islands may be formed by the gradual accumulation of the deposits of coral polyps, and even considerable parts of continents such as most of the peninsula

FIG. 113.—A group of sea anemones. (After Andres.)

FIG. 114.—A fringing reef with many varieties of corals. (After Saville-Kent.)

of Florida. The red coral, much used in making ornaments, is derived from a species which occurs in the Mediterranean Sea. Some forms allied to the true coral form branching colonies which secrete a skeleton of a tough, horny substance resembling chitin. These include the sea-mats, sea-fans and black corals, etc., many of which are commonly mistaken for seaweed. The graceful form and vivid colors of many of the colonies of coral

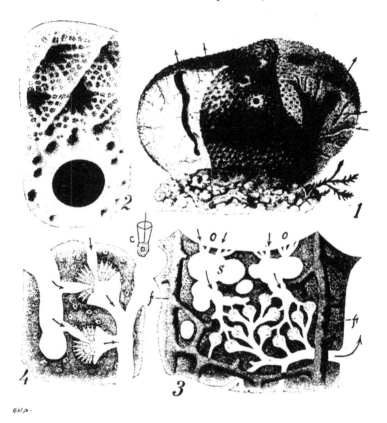

Fig. 115.—The common bath sponge. 1, shows the sponge with parts cut away to show the channels through which water flows in and out, the arrows indicating the directions of the flow; 2, a part of the surface highly magnified; 3, a section through the sponge showing the horny fibers, f, the incurrent orifices, o, and the chambers lined by ciliated cells; 4, ciliated chambers enlarged, a single collared cell at c. (From a Pfurtscheller wall chart.)

polyps often give to coral reefs the appearance of beautiful submarine gardens. Here thrive numerous other animals which seek the shelter afforded by masses of coral rock; many of these forms are also conspicuously colored and present a marked contrast to the transparency, or the inconspicuous colors, of most of the unprotected animals of the open sea.

The Cœlenterates known as Ctenophores, or comb-bearers have a fairly pronounced bilateral symmetry. They swim by means of eight rows of comb plates which act like so many small paddles. They have as a rule two long and very contractile tentacles armed with adhesive bodies which aid in the capture of prey. Nearly all of the Ctenophores are beautiful, transparent, jelly-like animals, and all of the species are confined to the sea.

The Porifera, or Sponges, were formerly regarded as vegetable growths, partly no doubt on account of their attachment and mode of growth, and partly because they show but a slight degree of activity. The structure of a sponge is best studied in one of the calcareous sponges such as Grantia. The body of Grantia is cylindrical with a central cavity which opens outward by a mouth, or *osculum*. The sides are perforated by pores leading to canals which open into the central cavity. Through these canals a current of water is carried from the outside to the central cavity, by means of the beating of flagella. The minute organisms carried by these currents supply the sponge with its food which is digested within the bodies of the cells lining the canals (intracellular digestion). The sponge has no organs of circulation or excretion, and no nervous system or sense organs. Like Hydra, the body has an inner layer, or entoderm, and an outer layer, or ectoderm; but between these are other cells some of which form the skeleton, or supporting tissue, of the body. In the calcareous sponges this consists of spicules of carbonate of lime. In other sponges the skeleton may be composed of silica as in the beautiful glass sponges.

In others it may be formed entirely or in part of a horny substance, as in our common bath sponges.

Sponges may build up large masses by budding; in this way they frequently give rise to detached individuals; but they also reproduce sexually by means of ova and spermatozoa. The eggs of many species produce free-swimming larvæ which finally settle down and develop into a small attached sponge. While most sponges are marine, there are a few fresh-water forms (Spongilla and allied genera). The sponges of commerce are fished up in certain localities, by long rakes, or by means of divers. The animal matter is allowed to decay, and the horny residue is bleached and cleaned before the sponge is ready for use.

CHAPTER XVI

THE PROTOZOA OR THE SIMPLEST ANIMALS

The lowest and simplest of all animals are the Protozoa. The group differs from the animals that have been studied in that the individual consists of but a single cell. Most of the Protozoa are of microscopic size, and some are so small that the highest powers of the microscope are necessary to detect them. The commonly recognized classes of the Protozoa may be separated by the following key:

A. Protozoa, at least at some period of life, moving by means of cilia. . . . . .Infusoria.
AA. Body devoid of cilia.
  B. Protozoa furnished with one or more flagella. . . .
  . . . . Flagellata.
BB. Body devoid of flagella.
  C. Usually free forms, with pseudopodia. . Sarcodina.
  CC. Exclusively parasitic forms, multiplying by means of spores and generally devoid of pseudopodia in the mature state. . . . . .Sporozoa.

Of these classes the Infusoria have the most complex organization. It is easy to obtain a typical infusorian in the common slipper animalcule, Paramœcium, which usually makes its appearance in infusions of hay or other vegetable matter. If one places a quantity of hay in some water coming from a pond or stream that contains more or less plant life, it is probable that Paramœcia will make their appearance in the mixture in the course of one or more weeks. They may be recognized by their uniformly ciliated body which has the general shape of a cigar with a broad oblique groove on one side. At the end of the oblique groove is a short gullet down which the food of Paramœcium is swept by the action of cilia. At the end of the gullet is a small enlargement, the crop, where the food accumulates forming a sort of ball. When the crop is filled with food it is pinched off by the contraction of the surrounding substance, and the mass of food with a small amount of water passes into the semi-fluid interior of the body. The small vesicles with their contained food are called food vacuoles. By a circular movement of the inner substance, or *endoplasm*, the food is slowly carried about the body, in the meantime undergoing a

Fig. 116.—Paramœcium. *cil*, cilia; *cv*, contractile vacuoles; *f*, food vacuoles; *g*, gullet with crop at end; *n*, macronucleus; *n'*, micronucleus.

process of digestion. The undigested residue is discharged at the surface of the body at a spot a little behind the mouth. Surplus water and products of excretion are got rid of through the hollow vesicles called *contractile vacuoles*, located near either end of the body. These may be seen to slowly swell from the accumulation of fluid and then suddenly contract, expelling the fluid to the outside of the body. Paramœcia eliminate through these organs several times their own bulk of water every day.

Paramœcium, like most other Infusoria, contains two kinds of nuclei, a large one, the *meganucleus*, and a small one, the *micronucleus* which usually lies close against the meganucleus. Reproduction in Paramœcium takes place by transverse fission. Both meganucleus and micronucleus divide and the two parts pass toward either end of the body which becomes constricted in the middle by a transverse furrow and finally pinches in two. Paramœcium may reproduce in this way for several hundred generations, but finally it is interrupted by another process which is called *conjugation*. In this process the Paramœcia come together in pairs and become united by their oral surfaces. When they are united complicated changes in the nuclei occur, during which the Paramœcia exchange

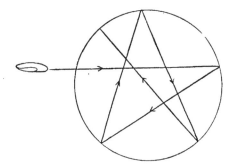

FIG. 117.—Diagram of the course of a Paramœcium in a drop of dilute acid surrounded by water.

a part of their nuclear material. After this they separate and continue dividing by fission until another conjugation period comes about. It is held by some biologists that conjugation regenerates, or puts new vitality into the stock.

The body of Paramœcium is asymmetrical, and as the animal swims through the water by the beating of the cilia it rotates on its long axis and describes a spiral path. When colliding with an object, or when stimulated in any other way, Paramœcium swims backward by reversing the stroke of the cilia, turns toward the side opposite the mouth, and then goes ahead. This reaction has been called the *motor reflex* and is performed by Paramœcium in response to all sorts of stimuli, in much the same way. In fact Paramœcium does little else except swim forward and give the motor reflex when it meets with a stimulating agent, so that its behavior is remarkably simple. Sometimes it tends to remain quiet with its oral side in contact with some solid object. Advantage is taken of this trait, by placing a bit of cotton wool in a drop of water containing Paramœcia, when we wish to keep the creature quiet for study.

Many infusoria have the property of secreting a coating, or *cyst*, about themselves, in which they are able to withstand conditions that would otherwise prove fatal. The cysts of some infusorians have been kept dry for over a year, when they gave rise to living Infusoria after being placed in water. If ponds dry up in summer the infusorians, and many other protozoa also, may go into an encysted state until the ponds become filled again. The dried cysts of protozoa may be blown for miles in the dust and thus scatter the species very widely. The great ease with which these minute forms become scattered accounts for the world wide distribution of many species.

The flagellate protozoa are devoid of cilia but they swim by means of one or more whip-like organs called *flagella*. Many of them have a mouth by which they take in food (holozoic forms) as the higher animals do; others imbibe food in soluble form through the body wall. Of the latter some live in decaying substances (saprophytic forms), some live within the bodies of other organisms (parasitic forms), and others live by the imbibition of inorganic substances (holophytic forms). One of the latter group, *Euglena viridis*, combines many characteristics of both plants and animals. This species is provided at the anterior end with a single flagellum inserted in a small mouth. Near the base of the flagellum is a small eye-spot. In the endoplasm there are numerous bodies containing *chlorophyll*, a compound which enables the green plants to utilize carbon dioxide in building up their living substance. Euglenas exposed to light take in CO_2 and give off O, just as the higher plants do. At the same time they swim about like animals and are said to take in a small amount of solid food through the mouth. Like Paramœcium, Euglena multiplies by fission, but it divides

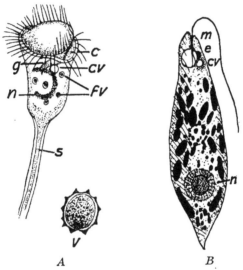

Fig. 118.—*A*, Vorticella; *c*, cilia; *cv*, contractile vacuole; *fv*, food vacuole; *g*, gullet; *n*, nucleus; *s*, contractile stalk; *v*, cyst of Vorticella; *B*, a flagellate *Euglena viridis.* *cv*, contractile vacuoles; *e*, eye spot; *m*, mouth; *n*, nucleus.

longitudinally as do most other flagellates, instead of transversely. At times it may go into an encysted state in which it sometimes divides into two or more individuals.

Some flagellates are more completely plant-like than Euglena, while there are others which are more like typical animals. In fact the animal and plant kingdoms seem to draw together in the flagellates which form a sort of common base from which they both diverge. The plant-like flagellates lead up to the simpler algæ and thence on to the higher plants; the animal-like flagellates lead on to other groups of animals. There is evidence that the flagellates are related to the simplest of all known organisms, the bacteria, which we may regard as standing at the very root of the tree of life.

Some of the flagellates that live within the bodies of animals are the causes of very severe diseases. Chief among these forms are the *trypanosomes* which are found generally in the blood of the infected animal. Nagana, which carries off thousands of horses in Africa, is caused by a species of trypanosome. The disease is conveyed by means of the bite of the tsetse-fly, much as malaria is carried from one person to another by the mosquito. One of the worst scourges of humanity that is known, the "sleeping sickness" of Africa, which is estimated to have carried off in the district of Uganda some one hundred thousand natives in four years, is caused by another trypanosome, *Trypanosoma gambiense*. The disease in its later stages is accompanied by extreme drowsiness which gave it its name, and it almost always results in the death of the patient. It is now known that this disease is conveyed from man to man by a species of tsetse-fly, but one that is different from the species that carries nagana.

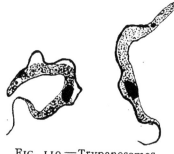

FIG. 119.—Trypanosomes.

In the group Sarcodina the body does not have a fixed outline, for it has the power of pushing out and withdrawing projections called *pseudopodia* (false feet) which serve both for locomotion and the capture of food. One of the simplest and best known of the group is the common *Amœba proteus*. The organism appears like a mass of animated jelly, with no constant form. The outer part of its body, the *ectoplasm*, is clearer and firmer than the inner part, or *endoplasm*, which is granular and more fluid. There is a single nucleus and a contractile vacuole. Amœba flows around its food and takes it into its endoplasm where it is digested in food vacuoles. Multiplication is commonly by fission, although after surrounding itself

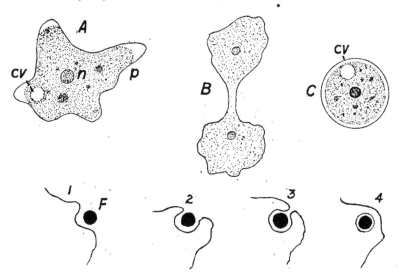

Fig. 120.—*Amœba proteus.* A, active state; *cv*, contractile vacuole; *n*, nucleus; *p*, pseudopod; B, Amœba dividing; C, cyst; 1-4, stages in ingesting a particle of food, F.

with a cyst Amœba may divide up into minute bodies called *spores*, which ultimately break out and become small Amœbæ.

The common Amœba is generally found in fresh waters and usually appears in cultures such as those used to obtain Paramœcium. There are several kinds of Amœba, a few of which are parasitic in the bodies of animals. Certain inflammatory diseases of the human intestine (amœbic dysentery) have been traced to Amœbas and related organisms. Many forms allied to Amœba live within a shell which is formed either by secretion or by the aggregation of foreign particles. One large group of these organisms, the Foraminifera, is abundantly represented in the sea, in certain parts of which extensive deposits are formed by the accumulation of their minute shells. Chalk is a deposit which is mainly formed of the shells of these animals. Another very large group, the Radiolaria, is confined to the sea. Many species have beautiful silicious skeletons and in some parts of the ocean there are extensive deposits formed from the remains of these animals.

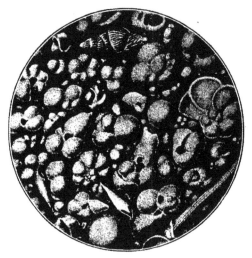

FIG. 121.—Shells of Foraminifera from the bottom of the Indian Ocean.

The Sporozoa, which, as the name implies, are characteristically spore-producing organisms, are all parasitic in the bodies of animals. One large division, the Gregarines, are parasitic in invertebrate animals, where their favorite situation is in the alimentary canal. Another group, the Coccidea, infest both vertebrates and invertebrates. The Hæmosporidia are blood parasites and include the common malaria parasites (Plasmodium) which have already been mentioned in treating of the mosquito. The organism that causes Texas fever in cattle and which is conveyed through the bites of wood ticks is another member of this group.

Many diseases of fishes are caused by other species of sporozoans. The disease called "pebrine," which attacks silk-worms and formerly caused very great losses to the silk industry of France, is caused by a sporozoan.

This is one of the comparatively few diseases which can be transmitted from parent to offspring through the egg. Louis Pasteur, a man famous for his valuable work

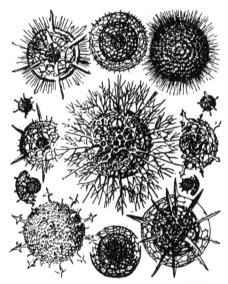

FIG. 122.—Skeletons of Radiolarians. (After Hæckel.)

in establishing the germ theory of disease, discovered that infected silk-worm moths could be distinguished from the others by microscopic examination of the blood. This fact made it possible, by rejecting the infected individuals, to check the spread of the disease, which has since caused much less damage.

Although small in size the protozoa by their vast numbers are important factors in the life of the world. As we have seen they produce many diseases in animals and man;

they have played an important part in building up certain deposits of the earth's crust; but most important of

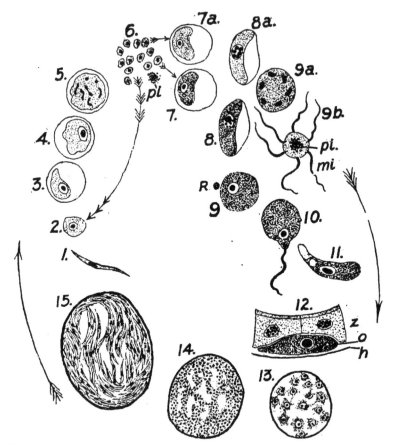

Fig. 123.—Life history of the malarial parasite. *l*, sporozoit as it enters the blood from the bite of a mosquito. This sporozoit becomes an amœboid body, 2, which enters a red blood corpuscle and grows, 3-5; 6, amœboid bodies into which the parasite breaks up and which enter new corpuscles and repeat the same history until the sexual cycle appears (7-11); 7-9, female cells; 7a-9b, male cells; 10, union of female cell with filamentous male cell; (this occurs in the stomach of a mosquito); 11, fertilized cell; 12, the same when imbedded in the wall of the stomach of the mosquito; 13-15, growth of this cell, multiplication of nuclei, and breaking up of protoplasm to form numerous spindle-shaped sporozoits, many of which later get into the salivary gland of the mosquito. (After Schaudinn.)

all is the rôle they play in the food relations of other organisms. Protozoa devour bacteria and other simple plants;

they prey upon other protozoa and even the smaller many-celled animals, and they get rid of disintegrating animal and vegetable matter of various kinds. In turn they are eaten by larger organisms. Together with the unicellular plants they afford most of the food of that large class of animals which, like the clams, sweep in their food supply by the action of cilia. They are eaten by hordes of smaller animals, such as copepods, free-swimming larvæ, etc. The smaller animals in turn supply food for larger animals, such as fishes, and it is proverbial that the big fishes eat up the little ones. The aquatic world would become a vast grave yard were it not for the unicellular plants and animals.

The protozoa are of much interest on account of their simplicity of structure and behavior. Animal life is here reduced to its lowest terms. Within a single cell is contained, as in germ, the power of performing most of the functions which are discharged by the various special organs of the bodies of higher animals. In the simple, almost structureless body of an Amœba, we have locomotion without limbs or permanent organs, digestion without stomach or intestine, respiration without lungs or gills, circulation without heart or blood vessels, contraction without muscles, and response to stimulation without sense organs or nervous system. The living substance of the body performs all these functions, not so readily as each would be performed by organs devoted solely to one particular activity, but still sufficiently well to enable the Amœba to make its living and propagate its kind. As we pass up the scale of life we find these various functions taken over by different organs which become perfected along one special line, at the same time losing the ability to do other things. This process is called the *physiological division of labor*, and it is quite analogous to the division of labor in human society. If everyone were a jack of all trades we should not be able to get on as well or have as many different things we want, as when different people make different articles and exchange their products. Society profits by specialized labor and there is every reason to believe that individual organisms do the same.

CHAPTER XVII

THE LOWEST VERTEBRATES AND THEIR NEAREST ALLIES

The various groups of animals thus far studied are collectively known as the *Invertebrates* on account of the absence of a vertebral column or back bone. We now pass to the *Vertebrates* in which a vertebral column is one of the most characteristic features of structure. In any vertebrate such as a fish, frog, bird, or horse we find that there are several fundamental characters which are very different from those prevailing among the invertebrates; the central nervous system is dorsal in position, the heart lies below (ventral to) the alimentary canal instead of above it; and the skeleton is an internal one, although in some vertebrates as in turtles, an outer skeleton may be present also.

FIG. 124.—A tunicate.

While most animals may be classed without hesitation as vertebrate or invertebrate there are a few of more or less intermediate position. Some of these are so different in appearance from the true vertebrates that their relationship to the latter would never be suspected upon ordinary observation. Such is the case, for instance, with the tunicates or "sea-squirts." Most of these animals are sac-like creatures living attached to rocks and sea-weed, and they derive the name "sea-squirt" from their habit of squirting out water when they are irritated;

the term tunicate refers to the usually tough tunic or covering which envelopes the body. When the development of these animals came to be studied, the surprising fact was revealed that the early embryonic stages strikingly resemble the corresponding stages of vertebrate animals, and it was found also that in the larval period of the tunicate there is a dorsal nerve cord, gill-slits which open from the pharynx to the outside like those of fishes,

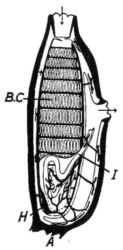

Fig. 125.—Internal structure of a tunicate. *A*, point of attachment; *BC*, branchial or gill chamber into which water enters through the orifice shown in the upper part of the figure. After passing through the numerous gill slits in the wall of this chamber the water is discharged through the orifice shown at the right of the figure; *H*, heart; *I*, intestine. (After Herdman.)

a ventral heart, and a notochord, or rod-like structure which corresponds to the primitive spinal column of the vertebrates. The tunicate larva is a free-swimming animal possessing the essential features of vertebrate structure. Sooner or later the larva settles down and becomes attached by its head; the tail is resorbed, and a complex metamorphosis ensues in which most of the vertebrate characters are either lost or much obscured. The adult tunicate is a degenerate animal, and were it not for our knowledge of its early stages its real relationship to the vertebrates would not be apparent. As in the barnacles, and especially such degenerate forms as Sacculina, the clue to the real affinities of these animals was first revealed through a study of development.

There is another group represented by Balanoglossus and a few related genera which also show relationships to the vertebrates. Balanoglossus is a worm-like animal which burrows in the mud of the sea bottom by means of a muscular proboscis. In the anterior part of the body there are a number of gill-slits which lead from the pharynx to the outside and serve for the exit of water taken in through the mouth. There is a dorsal nerve cord and a structure dorsal to the pharynx which is regarded as representing the notochord. Balanoglossus also resembles the vertebrates in many features of its early development, but its relationship is much less close than that of the tunicates.

In the lancelet, or Amphioxus, the vertebrate characters are much more conspicuous. This animal, which Professor Haeckel has called the most interesting vertebrate next to man, has a narrow, laterally flattened body tapering toward both ends. It lives in the sea partly buried in the sand in which it can burrow, when disturbed, with remarkable quickness. Like the tunicates its food consists of small bodies swept into the alimentary canal by means of cilia. The water that is taken into the mouth passes out through numerous fine gill-slits in the wall of the pharynx, while the solid particles are retained and swept by ciliary action into the intestine. The backbone is represented by a firm rod-shaped notochord extending above the alimentary canal. Dorsal to the notochord lies the nerve cord, the anterior end of which is slightly enlarged to form a sort of brain. This nerve cord corresponds to both the brain and spinal cord in ourselves; but how remarkably simple is the brain in this low creature compared with the brain of even a fish or a frog! A pigment spot in the wall of the brain marks the position of a rudimentary eye. The heart is represented by a tubular blood vessel on the ventral side which propels the blood forward; the blood then flows through the vessels in the gills, where respiration is effected and then backward in a dorsal vessel.

Although **Amphioxus** has no limbs or skull and but a suggestion of a brain its vertebrate characters are undoubted, and it matters little whether it is classed just within or just without the vertebrate group so long as we recognize its affinities. It is now customary to group the

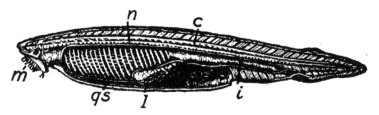

FIG. 126.—Amphioxus; *c*, nerve cord; *gs*, gill slits; *i*, intestine; *l*, liver; *m*, mouth; *n*, notochord.

vertebrates, **Amphioxus**, the tunicates, Balanoglossus and a few other forms in a large phylum called Chordata, the members of which all possess a notochord at some period of their life history. From what group of the invertebrates the chordates took their origin is a question to which the few intermediate groups which now exist do not enable us to give a certain answer.

The lowest true vertebrates, if we except Amphioxus, are found in the class of Cyclostomes, or round mouths, which include the lampreys and the hag-fishes. These animals have long eel-like bodies without any traces of limbs. There is a cartilaginous skull enclosing a well developed brain and a notochord which is not segmented. The gills lie in pouches, and the water passes out through one or more pairs of apertures in the sides of the body.

Most of the cyclostomes live upon other fishes to which they attach themselves by a sucker-like mouth which is furnished with a rasping apparatus for abrading the flesh. The lampreys are mostly fresh-water forms and the few marine species migrate up rivers to breed, the eggs of several species being deposited in rude nests constructed out of small stones. The hag-fishes are marine; often they bore through the body walls of fishes and devour most of the internal flesh.

Fig. 127.—A brook lamprey. (Modified from Gage.)

The vertebrates, in the most restricted sense of the term, comprise the generally recognized classes of Cyclostomes, Fishes, Amphibians or Batrachians, Reptiles, Birds and Mammals. Originally the vertebrates were aquatic animals breathing by means of gills, and the lowest classes, the cyclostomes and fishes, still retain the ancient habit of living in the water. The Amphibians, as the name implies (amphi, both and bios, life) live both in water and on land, the more fish-like members being aquatic gill-breathers, while the higher amphibians, such as frogs and toads, live upon land; but even these tend to remain near the water or at least in moist surroundings. The reptiles, birds and mammals have become primarily terrestrial animals.

CHAPTER XVIII

THE FISHES

In treating of fishes we shall describe first a typical fish such as a perch, sunfish, trout or bass, some one of which is easily obtainable in almost every locality. One notable feature of the organization of the fish is the adaptation of the form of its body for gliding through the water with the least amount of resistance. The posterior part of the body tapers into a thin, vertical tail fin which is expanded to give it a broad purchase against the water and it is strengthened by a number of long, flexible *rays*. The tail

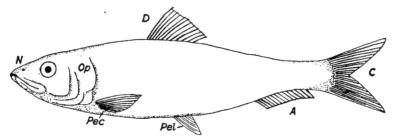

FIG. 128.—A typical fish. *A*, anal fin; *c*, caudal or tail fin; *D*, dorsal fin; *N*, nostril; *Op*, operculum; *Pec*, pectoral fin; *Pel*, pelvic fin.

is the principal organ of locomotion in the fish and the bulk of the muscle of the body is used in effecting the to and fro movement of this organ, which results in propelling the animal forward. As in most vertebrates a typical fish has four paired limbs represented by the anterior, or *pectoral*, and the posterior, or *pelvic*, fins. These, like the tail fins, are thin, flattened organs strengthened by means of rays. While the lateral or paired fins may aid in swimming forward, their main uses are to keep the body in a state of balance and to change the direction of movement. There are also median fins consisting of one or more dorsal fins above, and an anal fin on the lower side. In addition to the flexible and frequently branched rays, the fins may be strengthened by sharper and more rigid spines which serve also as organs of defense. Many fish erect these spines when angry or in danger.

The surface of the body is covered by hard scales which overlap so as to form a uniform protective layer. The scales are embedded in the skin and grow with the growth of the fish. Looked at through the microscope they show a series of concentric lines indicating successive stages of growth. In some kinds of fish (catfish) the scales may be absent, while in others, such as the gar pike, they may be large and very hard, forming a veritable coat of mail.

On the head of the fish the large eyes are noteworthy on account of the lack of eye-lids, so that they are permanently "open." The nostrils, of which there may be two pairs in some fishes, do not open into the mouth cavity as they do in higher vertebrates, but they lead to the organs of smell. At the side of the head is a large flap, the *operculum*, which covers over the gills. The jaws of the fish are quite different from our own in that the upper jaw is movable, instead of being firmly united to the skull. The jaws are furnished with sharp, conical teeth which are fitted for seizing and retaining prey; for the fish does not take time to chew its food, but swallows it entire. Teeth are frequently present also on the roof of the mouth, and in some fishes in other parts of the mouth cavity and even in the throat. Back of the mouth is the *pharynx*, the lateral walls of which are perforated by four pairs of *gill-slits*, between which lie the rows of slender, red filaments which constitute the gills. Water is taken into the mouth, passed through the gill-slits, bathing the gill filaments on its way, and then is forced out behind the edge of the operculum. The more or less regular movements of the jaws and operculum have to do with taking in water and forcing it out past the organs of respiration.

All of the fishes mentioned above have a bony skeleton composed of a large number of separate bones. There is a complex skull consisting of the cranium, or brain case, the jaw bones and their supports, and various other parts. Joined to the hind end of the skull is the vertebral column consisting of numerous bi-concave vertebræ; each vertebra has a dorsal arch, covering the spinal cord, and many of the anterior vertebræ are connected below with ribs which partially surround the body cavity. The rays of the pectoral and pelvic fins are joined to bony frameworks called respectively the *pectoral* and *pelvic arches*.

One of the most peculiar of the internal organs is the so-called swimming bladder, or "air bladder," which lies in the upper part of the body cavity. This organ is filled with gas secreted by the fish and serves as a sort of float. In some species this air bladder communicates by a duct with the esophagus. Among the lung fishes (Dipnoi) the walls of this body are well supplied with blood vessels and subserve the function of respiration.

Most of the bony fishes produce a large number of eggs. During the egg-laying period the females are usually accompanied by the males and when the eggs are extruded the males discharge their sperm, or "milt," over them and thus effect their fertilization. In the breeding season the males of many species, such as our common sunfishes, are more brilliantly colored than at other times, and display themselves before the females in a manner that suggests courtship similar to that practised by the males of many birds. It is a common practice among fishes to lay their eggs more or less indiscriminately and then, after they are fertilized, to leave them to their fate. Many marine fishes like the flounders lay eggs that float on the surface and are protected from their enemies by their remarkable transparency. Some species make nests for receiving the eggs; it is usually the male which performs this task.

In the common dogfish, or Amia, of our lakes and streams the male constructs a rude nest by pushing about some stones on the bottom and then induces a female to enter the nest where she deposits her eggs or, in the lan-

FIG. 129.—Showing the nest of a horned dace with the male and female fish on the nest. The stream flows in the direction indicated by the arrows. (After Reighard.)

guage of fishermen, "spawns." The male after fertilizing the eggs stands guard over them and rushes out to attack any other fish that ventures too near the sacred premises. Even after the eggs hatch, the male accompanies the young brood until they begin to scatter and shift for themselves. The male of the common stickleback constructs a more elaborate nest out of sticks and bits of grass and weeds. The males are irascible little creatures and defend the nest with much valor. In certain marine catfishes the male protects the eggs by carrying them in his mouth. Those fishes which simply lay their eggs in the water without further care are frequently compelled to produce enormous numbers of eggs to make good the great loss due to lack of protection. The forms which lay their eggs in nests or guard them after they are laid lay comparatively few eggs.

Some fishes make extensive journeys before depositing their eggs. The Columbia River salmon during the spring of the year leaves the ocean, where it spends a great part of its life, and entering the mouth of a river quickly swims up stream. In the beginning of their journey the fish are well fed and full of vigor; they require all their stored-up

Fig. 130.—Quinnat salmon. (From the Report of the Calif. Fish and Game Commission.)

supply of energy, since they take no food after entering the fresh water. When the salmon are "running," the water may be densely crowded with them, and the fishermen whose canneries line the banks of the Columbia River reap a rich harvest, for they have their nets spread for the unwary travellers and haul them in by hundreds of tons. Those fortunate enough to escape being made up into canned salmon, press on through rapids and often leap over low falls until they reach the smaller tributaries of the stream, in many cases over one thousand miles from its mouth. During their course the males become lean and battered, and they acquire a peculiar lengthening of the lower jaw and an increased development of teeth which are of value in their frequent combats with others of their own sex. When the fishes finally arrive at a suitable breeding place in some shallow stream, the eggs are laid and fertilized, after which the life of the parent fish is short. Thenceforth, the mission of their long and perilous journey accomplished, they live only in their posterity.

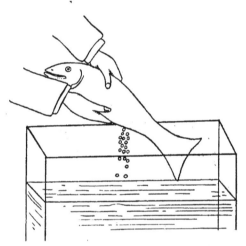

FIG. 131.—Stripping fish to obtain the eggs.

The young salmon gradually works down stream, growing in the meantime from its diet of worms, flies and other small creatures, and finally reaches the ocean where it lives until it in turn comes to obey the mysterious call to enter the river and sacrifice itself for the perpetuation of the species.

The migration of the common eel Anguilla is the reverse of that of the salmon, for the adults go down the rivers to breed in the ocean and the young migrate up the streams and live for most of their lives in fresh water.

As the eggs of nearly all food fishes are fertilized outside of the body, advantage is taken of this fact in the artificial propagation of many of the more valuable species. The United States government and various states support fish hatcheries where young fishes are reared and then let loose, to replenish, so far as possible, the numbers taken by fishermen. To obtain the eggs a ripe female is taken in the hands and "stripped" by slowly compressing the fish from before backward, thus forcing the eggs out of the body. The sperm, or milt, of the male, which is obtained in a similar manner, is mixed with the water containing the eggs and causes them to be fertilized. The eggs undergo

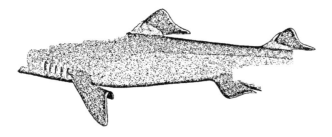

FIG. 132.—A shark, *Squalus acanthias*. (After Dean.)

their development in specially prepared receptacles and the young are set adrift. Through the fish commissions, various lakes and streams are stocked with desired species, sometimes by introducing the eggs, in other cases by transferring the mature fish.

Fishes may be divided into three sub-classes: the Elasmobranchs, or cartilaginous fishes; the Teleostomi, or bony fishes; and the Dipnoi, or lung fishes. In the elasmobranchs which are represented by sharks, skates, rays, etc., the skeleton is composed of cartilage, the gill-slits open directly to the outside instead of being covered by an operculum, and the tail fin is typically asymmetrical, or heterocercal. The mouth and usually the nostrils are situated on the ventral surface of the head, and the body is commonly covered with placoid scales which are peculiar in having a sort of prominence, or denticle situated upon a flattened base. Various transitions between these scales and teeth can be traced in some forms, so that we may regard teeth and scales as corresponding or homologous organs.

The elasmobranchs with a single exception are all marine and carnivorous. The sharks are, as a rule, active, preda-

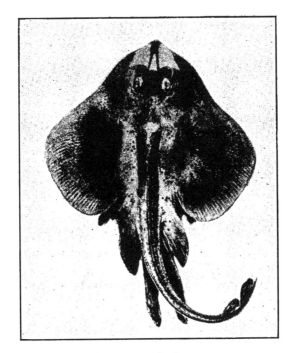

FIG. 133.—A skate.

tory animals feeding mostly upon other fishes. The whale shark Rhinodon may reach a length of 40–50 feet and the large, white, man-eating shark, Carcharias, 25 feet.

In the skates and rays the body is remarkably flattened, and adapted to living on the bottom where the animals feed mainly upon shell-fish and crustaceans. In the stingrays there is a pointed spine near the base of the tail, which is capable of inflicting a painful wound. The torpedo which is allied to the rays is remarkable in possessing a highly developed electric organ which may give rise to very severe electric shocks.

Most of the elasmobranchs lay large, yolk-laden eggs which are fertilized before they are laid. In many cases these hatch within the body of the mother so that the young are brought forth alive, but in some forms the eggs are enclosed in a horny shell, which is sometimes drawn

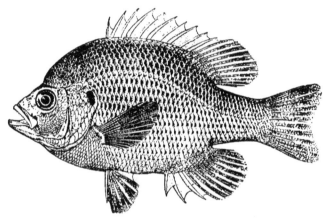

FIG. 134.—*Lepomis punctatus.* Chinquapin perch.

out at the two ends into four cords by means of which they become attached to sea-weeds.

The Teleostomi include the fishes with a more or less bony skeleton. The gill-slits are covered by an operculum and the body is generally covered with flattened scales. The most primitive of these are the ganoids which in early periods of the earth's history constituted a large and flourishing group. Now they are represented by a comparatively few forms, a large proportion of which are found in the fresh waters of North America. The large sturgeon of our lakes and rivers whose ovaries are sold as caviar; the slender, hard scaled gar pike; and the dog fish, Amia, are some of the better known representatives of this ancient group.

The largest of the two divisions of the bony fishes, the teleosts, includes the common fishes, such as cod, mackerel, perch, bass, minnows, catfish, eels, etc.—a vast and varied assemblage occurring in fresh water and in the sea, at all depths, and in all regions. They present almost every conceivable modification of structure consistent with remaining fishes, for what could be more diverse than the

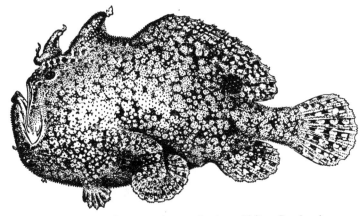

FIG. 135.—*Antennarius avalonis.* (After Jordan.)

puffy globe-fish, the grotesque sea-horse, the thread-like pipe-fish, the large headed "angler," and the almost impossible freaks of fish structure found in some of the denizens of the deep sea? Even a superficial treatment of these varied forms would require a whole volume and a large one at that.

The third sub-class of fishes, the Dipnoi or lung fishes, are represented, like the ganoids, by only a few scattered remnants of a once more numerous group. In these forms the air bladder communicates with the ventral side of the esophagus and functions as an organ for breathing air, although these fishes also breathe by means of gills. The air bladder of fishes is frequently regarded as homologous with the lungs of higher vertebrates, and in the Dipnoi

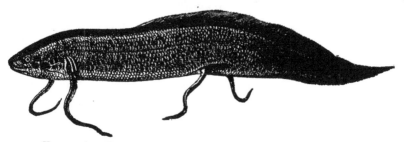

Fig. 136.—*Protopterus annectens*, lung fish. (From Boas.)

the air bladder approaches more closely to the structure and function of true lungs than in any other group of fishes.

Fig. 137.—Flying fish. (After Jordan.)

CHAPTER XIX

THE AMPHIBIA (FROGS, TOADS, NEWTS AND SALAMANDERS)

The Amphibians or Batrachians, as a class, show a certain proclivity for living in or near the water, and in almost all cases where species, as in frogs and toads, have taken to living on land they return to their ancient aquatic habitat to breed. In general the amphibians have a smooth skin devoid of scales or armor, and furnished with numerous mucous or slime glands which, in terrestrial species, serve to keep the skin constantly moist. The limbs in the amphibia are very different from those of fishes in that they are typically of the five-toed type and fitted for walking or leaping, like those of higher vertebrates.

The amphibians at present existing on the earth may be divided into three groups, the Urodeles or tailed amphibians (newts, salamanders, etc.), the Anura, or tailless forms (frogs, toads), and the Cœcilians which comprise a few legless worm-like species living in warm climates. In some of the lowest of the tailed amphibians we meet with many interesting fish-like characters such as the possession throughout life of external gills. These forms naturally live permanently in the water, although they are generally furnished also with lungs for breathing air. The mud-puppy, Necturus, is one of these primitive forms which is not uncommon in lakes and streams of the northeastern United States. Another is the peculiar *Proteus anguinus* which lives in dark caves in Austria. Like many other cave animals it has become entirely blind and the body has lost nearly all of its pigment.

Most of the higher tailed amphibians, such as the tritons, newts and salamanders, do not have gills in the adult state, although gills are usually present in the young. Some species live in the water and some on the land. Of the land forms the European spotted salamander is remarkable for its conspicuous colors. The skin of this animal secretes a very poisonous, milky fluid which affords it a sufficient protection, since most animals which live in the same region know that the spotted salamander is something to be left alone. Its conspicuous colors are supposed to be of value in enabling it to be easily recognized and therefore in preventing it from being molested. Such colors are commonly called "warning colors" and they are not infrequent in animals which are poisonous or distasteful. Some of the Urodeles, such as Triton, can regenerate missing legs or tail, or even the eye, but in the frogs and toads this power is very limited.

The tailless amphibians, or Anura, comprise the most highly developed amphibians. They are generally found on the land, but in most cases they lay their eggs in the water and the young hatch as tadpoles which resemble the most primitive urodeles in having external gills and well-developed tails. The series of forms which we meet in passing from the lowest to the highest Amphibia is roughly similar to the stages passed through in the development of an individual frog or toad.

The breeding season of frogs and toads is in the spring when the animals repair to the water to deposit their eggs. The eggs laid are surrounded by a transparent jelly which affords them a certain protection. During the breeding season the females are clasped by the males which eject their sperm over the egg masses as soon as they are extruded from the body of the female. The eggs once laid and fertilized, the frogs leave the water and resume their active predatory life. The long months of hibernation, when life was supported only by the food materials stored up in the tissues, leave the frogs lean and hungry, especially after the additional burden of maturing the reproductive cells. Life in the winter except in warm climates is spent in a dormant state, when the temperature of the body runs down and the vital activities become very sluggish. Frogs which are partially frozen until their legs are so brittle that they can be broken like icicles may subsequently revive, provided they are very slowly thawed out; if, however, they are frozen solid throughout they never regain life.

It is in the spring that the frogs make the most music. Ordinarily it is the males that do the croaking and it is supposed that the voice serves as a call, like the chirping of the male cricket, for bringing the sexes together. In some frogs the males have a pair of vocal sacs opening into the throat, which become inflated during the act of croaking. In others, as in the tree frogs, the production of sound is accompanied by a distension of the floor of the throat. The male toad produces only a relatively faint but peculiarly musical trill.

Frogs and toads are not particularly choice of what they eat, as they devour all sorts of insects, worms and other small creatures, but they are very peculiar in their methods of food taking. They are furnished with an extensile tongue which is joined to the front of the lower jaw and is capable of being thrust out of the mouth and withdrawn again with great quickness. Should an insect or worm be moving near by, the tongue may be shot out and the prey drawn back into the mouth, and quickly swallowed. It is only moving objects that attract frogs or toads, for they will go hungry in the midst of plenty unless some motion induces them to respond.

Frogs never drink as higher animals do but they obtain water by absorption through the skin. They lose water very rapidly by evaporation when kept in a dry atmosphere and present a much shrivelled appearance, but if placed in water again they soon become plump. Dryness is soon fatal to frogs and toads and they consequently rarely venture far from water, or at least they seek a moist retreat. The skin of these animals is an important organ of respiration as well as of absorption. While under ordinary circumstances respiration is carried on by both lungs and skin, skin respiration alone may suffice to maintain life if the animal is at a low temperature. Frogs often bury themselves in the mud at the bottom of ponds during the winter when respiration is naturally carried on through the skin alone.

Frogs (Ranidæ) are found in nearly all countries of the globe except in the colder latitudes where they cannot escape being frozen in the winter. The most common of our many North American species is the leopard frog, *Rana pipiens*, which ranges over a large part of the middle and eastern sections of the United States. Our largest species is the bull-frog, *Rana catesbiana* whose very hoarse croak resembles the roaring of a bull. It is generally found in or near water and has been sought so much for food that it has been very much reduced in numbers and is practically extinct in many regions where it was once abundant.

Toads (Bufonidæ) are generally more terrestrial in habit than frogs. The rough warty appearance of their skin is due to the development of large poison glands whose secretion is quite irritating to sensitive surfaces.

There is no foundation, however, for the superstition that handling toads produces warts, for they can be handled with perfect safety. Few animals are more useful to the farmer or gardener than these humble creatures for they devour large quantities of injurious insects which they catch during their nocturnal wanderings. Kirkland has estimated that in a farming section in Massachusetts

FIG. 138.—The bullfrog. (After Needham.)

every toad is worth several dollars on account of the cutworms alone, which it devours in a single season. For some reason many persons indulge in the repulsive proclivity of killing all the toads they meet with. Such conduct is not only foolish and cruel, but it is quite opposed to their own interests. The ugliness of the toad doubtless tends to make people treat it with contempt but like the homeliness of people whom we come to like, it becomes transformed upon closer acquaintance into a source of positive pleasure. Hodge remarks, "I pick up a toad a hundred times a season just to enjoy looking at its eye, a living, sparkling, ever-changing jewel, and his music in the springtime brings a pleasure that nothing else affords." Toads are easily kept in confinement and make interesting pets.

The tree frogs (Hylidæ) comprise an interesting family of rather small frogs most of which live a large part of

FIG. 139.—A toad, *Bufo halephilus*. (From photo by Holliger.)

their lives upon trees. The toes of most species are tipped with sucker-like adhesive pads which enable them to climb up vertical surfaces. The family in general is remarkable for the extensive changes of color which take place in response to the environment. *Hyla versicolor* is usually of a bright green color when among green leaves, a dull gray or brown when resting upon bark, and various intermediate shades under other conditions. This ability to change color is to a considerable extent protective, and is affected by light, temperature, rough or smooth contact, and a variety of other agencies. Although not commonly seen the tree frogs are very frequently heard, as the males are capable of making a noise which seems absurdly out of proportion to their diminutive bodies. Their song is more apt to be made in a moist atmosphere and this probably accounts for the fact that it is commonly regarded as prophetic of rain.

Nearly all the tailless amphibians undergo a metamorphosis, the early stages of which are passed in the water. In our common frogs and toads the larva, or tadpole, as it emerges from the jelly in which it has passed its embryonic development, is furnished with three pairs of external gills and a flattened tail by which it swims through the water much after the fashion of a fish. The young tadpole lives mainly on aquatic plants, although it may eat animal food also when occasion offers. As the tadpole grows, legs bud out; first the hinder pair and later the anterior ones. During the development of the lungs, the gills gradually disappear and the tadpole frequently comes to the surface for air. With the growth of the legs the tail becomes shorter and is finally resorbed into the body. While these changes are going on the young frog or toad, as we may now call it, gradually comes to move about on the land. Four or five years are required for our common species of frogs to become sufficiently mature to produce young, and they may live four or five years longer if they are fortunate enough to be spared from their many enemies.

CHAPTER XX

THE REPTILES

The reptiles, although commonly associated in the popular mind with the Amphibia, nevertheless constitute a very distinct class which is really more closely related to the birds than to any other group of vertebrate animals. The reptiles are lung breathers at all periods of their life and never have any gills, even in the young state. The body is covered with scales or encased in a bony armor. Living reptiles fall into the groups commonly designated as Ophidia (snakes), Lacertilia (lizards), Crocodilia

FIG. 140.—A garter snake. (After Van Denburgh and Slevin.)

(crocodiles and alligators), and Chelonia (turtles and tortoises).

In the snakes the body has become greatly elongated and very muscular. No limbs are present in most snakes, but in the pythons and boas there are rudiments of hind limbs and the pelvic girdle (see **Fig.** 235). Snakes are regarded as in some respects degenerate animals which have lost the limbs possessed by their ancestors. But while they may have lost certain organs they have developed a remarkable degree of strength, quickness and effectiveness which has enabled them to become one of the dominant groups of reptiles. By the winding movements of the body, snakes can progress with considerable rapidity, and they are especially adapted to making headway through masses of vegetation which would greatly impede the movements of other animals. The large overlapping scales of the ventral side of the body with their free posterior edges serve to facilitate forward movement by catching in the irregularities of the surface over which the animal glides.

Snakes are carnivorous and feed upon living animals. The teeth of snakes are conical and adapted for seizing and retaining prey which is always swallowed entire. The jaws are especially adapted to swallowing large animals in being separable from the skull at the base, thus permitting a great enlargement of the throat. Snakes frequently swallow animals whose bodies are much thicker than their own, and when distended with food they may remain for several days in a dormant condition while their meal is undergoing digestion.

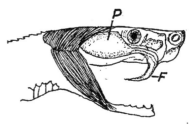

FIG. 141.—Jaws of rattlesnake showing fang, F, and poison sac, P. (After Weir Mitchell.)

Periodically snakes shed their outer skin which usually comes off in a single piece. Even the covering of the eyes is shed along with the rest, these organs being protected by a transparent part of the general skin instead of eyelids. While most snakes lay eggs, others, such as our common garter-snakes, bring forth their young alive. Despite stories of snakes taking their young into the mouth to protect them from danger there is almost no care for the young on the part of the parents; the young are left to shift for themselves which they are abundantly able to do upon their first appearance on the stage of life. Snakes in general are useful animals since they destroy large numbers of injurious insects; they are also enemies of mice, gophers, and other destructive small mammals. The bite of most species is scarcely painful, although there are several forms which have a well-deserved reputation for being dangerous. The most common of the venomous snakes of North America are the rattlesnakes which are provided with a jointed, horny rattle at the end of the tail, by which they make their peculiar rattling sound, when disturbed. This sound serves as a warning which all creatures that know what it means are only too ready to heed; it may be a service to the snake also in frightening away larger animals that might otherwise trample upon it. The poison of the rattlesnake is secreted by two large glands whose ducts lead to the base of a pair of large perforated fangs through which the poison is injected into the bite.

The bite of the rattlesnake has frequently resulted in death, and in all cases it should receive prompt treatment. A tight bandage should first be applied between the wound and the heart so as to check the return flow of blood in the veins, and the wound should be induced to bleed freely. Blood should be sucked from the wound and permanganate of potash should be administered either by injecting a solution into the wound or by rubbing in the crystals. Brandy and whisky have been much over-rated as remedies for snake bites, although they may be of value in stimulating the heart action, for snake venom acts as a heart poison.

The only poisonous snake in the United States, except those belonging to the rattlesnake family is the coral snake of the South, which is black with seventeen rings of red, bordered with yellow. In the old world, although there are no members of the rattlesnake family, there are many other snakes which are very dangerous. The cobra of India is responsible for the death of about twenty thousand persons a year.

The copperhead and the water moccasin are members of the same family as the rattlesnakes, but they have no rattle. The former is copper colored and lives mainly in the mountain districts of the Southern States; the latter is aquatic and, like the copperhead, is very poisonous.

Some snakes reach a great size, such as the boa constrictor, anaconda and the pythons some of which attain a length of thirty feet. These, while non-poisonous, swallow large animals after they have coiled about them and crushed them in their coils.

The lizards are mainly inhabitants of warm climates and lovers of dry places. There are very few species in the northern or eastern states, but in the arid regions of the southwest they are quite abundant. As a rule lizards are very active creatures, running over the ground and up trees with surprising quickness. On hot days the lizard seems especially to enjoy life, and he basks in the sunshine ever ready to dart at some insect that happens to move in his vicinity. In most lizards the long slender tail breaks off on slight provocation and when an individual is seized by this organ it is usually left in the hands of the enemy, while the animal makes its escape; a new tail is later regenerated.

The chameleons of the old world are lizards which are capable of striking and rapid changes of color. Commonly green, they may change to brown and various intermediate shades owing to the modifications of the pigment cells of the skin, as in the tree frogs. They are frequently kept as pets. The "horned toads" of the western states are lizards with flattened bodies and short tails. Their color resembles that of the soil. They readily live in captivity if fed upon the living insects. One of the largest of North American lizards is the Gila (pronounced heela) monster which is found in Arizona, New Mexico and Mexico. It is heavy, stocky animal and, unlike most lizards, usually sluggish in its habits. It is the only North American lizard whose bite is poisonous, the venom being conveyed to the wound by grooves in a pair of large teeth.

There are some lizards which have lost their legs, like the Cœcilians among the Amphibia, and they are frequently

FIG. 142.—Horned toad, *Phrynosoma blainvillei*. (After Bryant.)

therefore mistaken for snakes. Such is the case with the so-called joint-snake or glass-snake which receives its name from the fact that its tail is readily broken into fragments. This is because it is a lizard and not a true snake. There is a prevalent myth that the glass snake gathers together the joints of its tail and becomes whole again, but it is perhaps needless to say that the story is without foundation in fact.

The Crocodilians have the appearance of immense lizards, although they are quite different from the lizards in structure. In the United States they are represented by alligators of the southern rivers and by a species of true crocodile found in southern Florida. They spend most of the time in the water where they lie in wait for prey with the nostrils exposed at the surface. They often come out upon the banks to bask in the sunshine but they are comparatively awkward upon the land. Their eggs are laid in the sand and hatch out by the heat of the sun. Crocodiles occur in the Nile and other rivers of Africa and a related form, the gavial, inhabits the Ganges. Both alligators and crocodiles live mainly on fish, but they sometimes overcome fairly large animals which come to the water to drink. The American species have a wholesome fear of man and rightly so, since thousands of them are killed every year for their hides.

The most highly modified of existing reptiles are the Chelonians. The body of most turtles is enclosed in an armor of plates joined to the ribs and the backbone. The dorsal piece, or carapace, is composed in part of bony plates and in part of large horny scales overlying the plates, but not corresponding to them in position. The ventral piece, or plastron, is firmly joined to the carapace at the sides. The head and legs may be more or less completely withdrawn into the shell, and in the box turtles the plastron is formed of two movable plates united by a hinge joint which permits the two parts to be drawn up against the carapace so as to completely enclose the animal. The jaws of chelonians are entirely devoid of teeth, but they are furnished with a sharp, horny rim, by which they can retain hold of prey as well as inflict a severe bite. Most species are carnivorous, but there are several that feed upon vegetation. Among these are several species of land-tortoises which live entirely upon the land. One species occurs in the desert regions of Arizona and California and another in Texas and New Mexico. The gopher tortoise of the southern states is related to the preceding and derives its name from its habit of making long burrows in the sand. There are many species of turtles in the ponds and streams of the United States. One of the largest is the snapping turtle which sometimes reaches a weight of forty pounds. It is named from its habit of quickly snapping against an object of attack. According to Dugmore the "amputation of a finger by a medium-sized specimen, or a hand by a very large individual would be an accomplishment of no difficulty to the reptile."

In the sea-turtles the limbs are in the form of flippers adapted for swimming. The large leather-back turtle of the Atlantic may reach a length of six feet and a weight of a thousand pounds. The green turtle, and to a less extent, the loggerhead, are much sought after for food. The valuable tortoise shell of commerce is derived from another large marine species, the hawk's-bill turtle, which is widely distributed in the warmer seas. The eggs of turtles and tortoises are oblong and encased in a calcareous shell. They are usually buried in the sand near the water, and are hatched by the warmth of the sun.

CHAPTER XXI

THE BIRDS

The birds are so sharply distinguished from all other vertebrate animals that no one would make a mistake in assigning to its proper class even the most aberrant member of the group. There are other vertebrates that are able to fly, such as the bats among mammals and, in former periods of the world's history, some outlandish looking reptiles called pterodactyls; but the resemblance of these creatures to birds, aside from the possession of wings, is quite remote. All birds are furnished with feathers; they all have a horny bill; and they all have but a single pair of legs which are used for walking, hopping or running. This pair corresponds to the hind limbs of other vertebrates, the fore limbs being modified to form the wings.

The aerial life of birds has been one of the chief causes of their distinctive peculiarities of structure. Flight implies strength of bone and muscle, and expanse of surface for beating against the air, or for steering a course through it. The expanse of surface is mainly afforded by the feathers. While feathers are found in no creatures except birds they represent highly modified scales such as cover the bodies of reptiles. Nature is continually adapting old organs to new functions and, in evolving the feather from the scale, she has perfected a wonderfully complex and beautiful structure that seems at first to have little in common with the original source. Like scales, feathers are derived from small papillæ in the skin. Commonly birds shed or molt their feathers in the fall, but some birds molt at other times also. Frequently the plumage that replaces an older one is of different color and some birds have a regular alternation of summer and winter plumage. The ptarmigan, for instance, is brown and white in summer, but after the molt in the fall it takes on a coat of pure white feathers. In nearly all birds the first feathers that appear in the young are very different from those that come later. Sometimes as in young

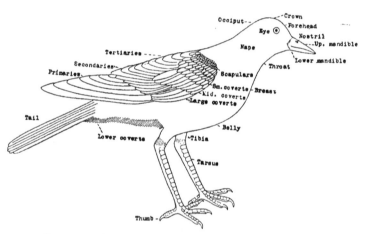

Fig. 143.—Diagram of a bird showing external parts. (Drawn by Miss H. M. Gilkey.)

chickens the plumage is in the form of a soft coat of down which resembles fine hairs, although it is not composed of hairs but of true feathers. Pin feathers are immature stages in the development of the plumage in which the feather is still surrounded by a sort of sac.

One function of feathers, like that of the fur of mammals, is to protect the body from cold and wet. The feathers on the wings and tail, however, which are much larger and stronger than those elsewhere, are used as organs of flight. The tail which can usually be spread out and contracted again like a fan is employed, like a rudder, as an organ of steering. Notice the movements of the tail as a bird changes its course or alights and you will see how this organ is used to guide the bird through the air. Feathers shed water easily; and they are aided in doing so by being kept more or less oily. There is an oil gland situated just over the base of the tail, and birds often take some of the oily secretion of this organ into the bill and distribute it over the plumage. Birds often preen their feathers, or set them in order, by working over them with the bill when they become disarranged.

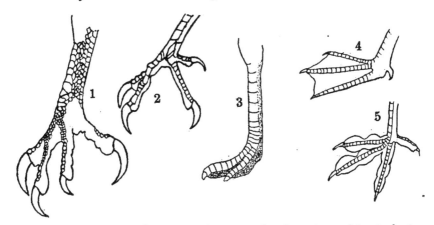

FIG. 144.—Feet of birds. 1, hawk; 2, woodpecker; 3, ostrich; 4, duck; 5, grebe.

The feet of birds are covered with tough horny scales. They have but four toes, one of which (and in some cases two) is generally directed backward. The toes end in a claw, or nail, which varies in shape according to the habits of birds. Many birds which swim or wade have the three front toes connected by a membrane, or web; or else the sides of the toes are furnished with flattened lobes. In birds of prey, such as hawks, owls and eagles, the feet are powerful and furnished with strong, curved claws which adapt them for seizing prey. In many climbing birds, such as the woodpeckers, there are two toes in front and two behind. Birds which perch generally have toes which automatically close up as the bird settles down upon a limb or perch.

The jaws of all living species of birds are entirely devoid of teeth and constitute what is known as the beak or bill; this has a tough, horny covering, and varies greatly in shape in different species. Commonly the bill is more or less conical and sharp at the tip, which adapts it for picking up seeds or insects. In birds of prey the bill is curved

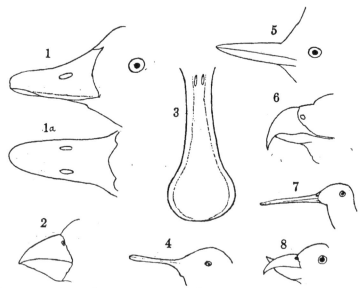

FIG. 145.—Bills of birds. 1, duck; 1a, same from above; 2, grosbeak; 3, spoonbill; 4, snipe; 5, kingfisher; 6, hawk; 7, humming bird; 8 crossbill.

downward in front, forming a sort of hook which is used in tearing the flesh of their victims. Ducks, geese and their allies have a flattened bill adapted for scooping up objects in the water or mud. The woodpecker's bill is strong and sharp like a chisel and is used for pecking holes in trees, which is accomplished by a rapid series of strikes the sound of which is familiar to everyone who has been long in the woods. The narrow, slender bill of the humming bird which is used to probe the necks of flowers for small insects or honey represents another type which is strongly contrasted with the short bill of night hawks and swallows, adapted for catching insects on the wing. A curious modification is seen in the crossbills in which the tips of the jaws are curved and crossed. This arrangement is peculiarly adapted for extracting from open pine cones the seeds upon which the crossbills feed. The largest

FIG. 146.—The toucan.

bills among birds occur in the toucans, or horn-bills, which live in South America.

As birds have no teeth they do not chew their food; while the objects may be broken up more or less by the bill they are usually swallowed whole. The function of grinding up food is performed by the gizzard which is a very strong muscular division of the alimentary canal. Usually birds that live more or less upon hard seeds swallow a quantity of gravel which aids in the grinding process. In many birds the esophagus expands below the base of the neck into a thin-walled crop in which a quantity of food is carried and then gradually passed back to the gizzard. After being ground up in the gizzard and partly digested there the food passes into the intestine where digestion is completed and the digested materials absorbed.

Not only do birds have well-developed lungs, but there are in most species, extensive air sacs connected with the lungs and extending into various parts of the body. In most birds many of the bones are hollow and contain prolongations of these air sacs. Air may therefore be carried to different parts of the body in a way that suggests a comparison with its distribution through the tracheal tubes of insects. A sparrow with a broken wing may even take air into its lungs through the hollow of its wing bones.

Birds are warm-blooded animals, and they have a rapid respiration which is greatly facilitated by the large surface afforded by the lungs and air sacs. They have a four-chambered heart and a complete double circulation such as occurs in ourselves.

The skeleton and muscular system of birds have become highly modified in relation to flight. To progress rapidly through the air means that there must be a large amount of muscle for moving the wings and accordingly we find the pectoral muscles, those extending from the breast bone, or sternum, to the wings, enormously developed. And to give adequate attachment for these muscles the sternum is not only of large size, but in all except a very few kinds of birds, it is furnished below with a large median ridge, or *keel*. The bones of the wings conform to the same general plan of structure as do those of the legs. In the outer part of the wing, or what corresponds to the hand in ourselves, some bones of the digits have been lost and others are fused together so that the fundamental plan is somewhat obscured. Such changes may be regarded as a natural consequence of modifying a fore leg so as to adapt it to the new function of flying.

All species of birds lay eggs, and with rare exceptions

birds sit upon their eggs or incubate them, their development being dependent upon the warmth afforded by the bird's body. Birds as a rule devote an unusual amount of care to the rearing of offspring and they afford many striking and attractive exhibitions of fidelity and devotion in their family life. The behavior of birds varies greatly,

FIG. 147.—A rookery of nesting birds on the Farallone Islands. (From a group in the museum of the California Academy of Sciences.)

however, in this regard. Among the lower, or more primitive, birds the eggs are laid either in simple, crude nests, or upon bare rocks or soil (see Fig. 147). The labor of incubation in these cases falls entirely upon the female as the associations of the sexes are very temporary and thus stand in marked contrast to the matings which occur in the higher forms. The young of the primitive birds when first hatched are generally active and require little attention from their parents. Young ducklings, for instance, will swim in the water, pick up food, flee from their enemies, and perform many other acts on the first day after being hatched. Lloyd Morgan tells of a young moor hen which swam almost as soon as it hatched out of the egg, and dived into the water as readily as an older bird. Very young chicks have the instinct to peck

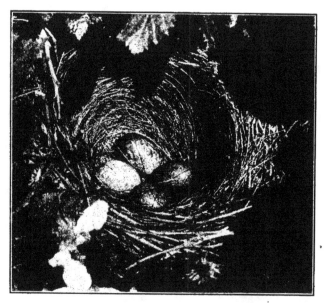

FIG. 148.—Nest and eggs of Brewer's blackbird. (From photo by Holliger.)

at small objects of about a certain size, but they quickly learn to avoid things with a disagreeable taste. They instinctively respond to the note of the mother hen that warns them of danger and rush to the mother or else hide. They also give a note themselves, the danger chirr, when any object causes them to become afraid and this note serves as a warning to the other chicks. One often sees them approaching an object such as a large bug of which they are half afraid and, after looking it over suspiciously, giving the danger chirr which means "beware!" When one chick musters up courage to peck at an object others usually follow its example. If the chick picks up a worm and bolts off with it the other chicks frequently take after the successful one in the endeavor to share its prize. All these acts and also many others are performed very soon after the chicks emerge from the egg. The young chick has a number of instincts which equip it, without previous experience, for most of the circumstances of its life. Hudson relates how the young of some birds will instinctively respond to the parent's call even before they break out of the egg shell. Birds which are active as soon as hatched flock about the mother bird who hunts food for them and gives them a certain protection. By imitating many of the actions of the parents the young learn to avoid enemies and derive many other advantages from their parents' experience.

In the higher birds, such as the song birds, the nest is built usually of small sticks, twigs and bits of grass and lined with down and other soft materials; and the young which are hatched in a weak and helpless state are fed and tended by their parents until ready to take flight. The common robin, for instance, which is a familiar visitor in the early spring, builds a nest usually in the bough of a tree, and both the male and female birds take turns in sitting upon the eggs which are hatched in about three weeks. The young, of which there are usually from two to five, remain in the nest until they acquire a coat of feathers (for they have but a scanty coat of pin feathers at first) and are then induced by their parents to leave, if they do not do so of their own accord. Herrick in his book on the "Home Life of Wild Birds" describes as follows the behavior of a family of robins whose nest he had carried, together with the bough of the tree on which it was built, to a convenient point for observation. "In exactly fifty-five minutes from the beginning of operations the mother appeared with a large grasshopper, which she gave to the young, and afterward cleaned the nest. The male came also, when the comparative safety of the new conditions had become apparent, but appeared with more caution. At first both birds flew to the tree by their accustomed paths and examined the place where the bough

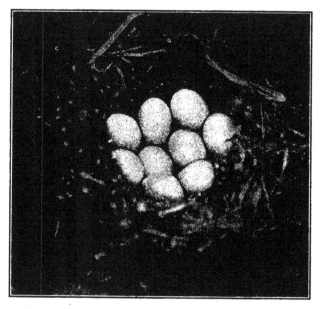

FIG. 149.—Nest and eggs of Massachusetts black duck. (After Forbush.)

had been lopped off, and in their admirable and fearless manner blustered about for a while, taking no pains to conceal their anger . . . . . The absence of sound in strange objects is alway reassuring and now Mother Robin could be seen perched on the top of an apple tree, surveying the field. She called *seet! seet!* while the grasshopper in her bill squirmed to get free, and the young chirped loudly in reply . . . . . Suddenly down comes one of the old birds with all its weight on the limb. The young have felt similar vibrations before and know what to expect. Up go three heads at once, each mounted on a slender stalk, and each bearing at its apex what might suggest a full blown, brilliant flower, for, as is well known, the extent of their gape is extraordinary and the inside of the mouth has a bright orange hue. The young tremble with violent emotions as they jostle, struggle, and call with undiminished zeal even after being fed."

"After the first visit had proved successful, confidence was established at once, the female and later the male

FIG. 150.—Robin catching an earthworm.

coming to the young at intervals of about five minutes, bringing grasshoppers, and occasionally removing the excreta . . . . . They frequently carried five or six insects at each load, when their bills would suggest a solid load of grasshoppers, all struggling to get free."

After feeding the young the parents carefully inspected the nest and freed it of any uncleanliness. "Then after inspection is over they fly to the nearest perch, and make haste to clean their bills and set their dress in order . . . . . One robin at the age of eleven days left the family circle early on August 13th, and at nine o'clock the two which remained were standing up and flopping their wings.

The old birds would come near, displaying tempting morsels in their bills but with no intention of feeding their young so long as they remained in the nest. By such tantalizing methods they soon drew them away. Both old and young hung about the apple tree for several days, when they disappeared and were not seen again."

Even after the young leave the nest they are accompanied by the parents for quite a while, and it is not uncommon to see a young robin two-thirds grown begging its indulgent parents for food, and being fed with angle-worms, when it is quite able to forage for itself.

The mating habits of birds are subject to great variation. In the more primitive species there is, as a rule, no permanent union of the sexes, the males and females separating after the breeding season is over. Many birds are polygamous, such as our domestic fowl, a single large, strong male going about with a flock of females, and driving away all weaker rivals. In most of the song birds, however, there is a more permanent union of the sexes, in some cases lasting until the death of one of the members of the pair. The males frequently take turns with the females in sitting upon the eggs, an office which the rooster among our domestic poultry would never condescend to perform. The males of many song birds also help in bringing food to the young, in cleaning the nest, and in some cases in bringing food to the female while she is incubating the eggs.

Among the higher birds, nesting is usually preceded by courtship, a ceremony which is dispensed with among the cruder and less gallant males of the lower birds.

It is very common for the males to be distinguished from the females by more brilliant and beautiful plumage and superior powers of song. Both of these characters are brought into play by the males who attempt to display themselves to the best advantage before the eyes of their intended mates. Darwin attempted to account for the superior qualities of the male birds by his theory of sexual selection, according to which the males that were the most brilliantly colored, or which sang most sweetly, or otherwise displayed themselves to the best advantage

FIG. 151.—Side view of male Argus pheasant, whilst displaying before female. (After Darwin.)

would be most likely to be chosen by the females as mates. Hence if this selection were continued generation after generation, the males would gradually be improved in respect to those qualities that appealed most strongly to the sensibilities of the female birds. Almost everyone has observed the strutting of the turkey gobbler with his erected feathers and expanded tail, and many are doubtless familiar with the much more beautiful display of the male peacock as he spreads out his magnificent, wonderfully marked tail feathers before the gaze of the presumably admiring pea hens. Darwin relates that during courtship "the bull-finch makes his advances in front of the female, and then puffs out his breast, so that many more of the crimson feathers are seen at once then otherwise would be the case. At the same time he twists and bows his black tail from side to side in a ludicrous manner. The male chaffinch also stands in front of the female, thus showing its red breast and 'blue bell' as the fanciers call his head . . . . . The common linnet distends his rosy breast, slightly expands his brown wings and tail, so as to make the best of them by exhibiting their white edgings."

As Darwin remarks "there is an intense degree of rivalry between males in their singing. Bird fanciers match their birds to see which will sing longest." Singing is most common during the breeding season. Many male birds which are not at all musical give utterances to cries and other noises during this season, which possibly serves rather to advertize their presence than to charm their hearers. The gabbling of the strutting male turkey, the harsh screaming of male parrots, and the hoarse cawing of male crows and rooks are certainly not musical to us, however they may appeal to the female bird.

It is well established that female birds often manifest a decided preference for certain males. Audubon states that female turkeys prefer the males of wild turkeys to those of their own domestic breed. There are several cases in which females have rejected their mates after they had lost their brilliant tail feathers or become otherwise mutilated, and female pigeons sometimes desert their own mates and take up with other males. How far an appreciation of beauty occurs in birds it is difficult to say.

Crows and magpies often carry away bright and colored objects. Some birds weave colored feathers into their nests, but the most remarkable exhibition of fondness for colored objects occurs in the bower birds of Australia. These birds erect a bower or tent-like structure, built of sticks and leaves. Around the entrance to the bower,

FIG. 152.—Humming birds, *Spathura underwoodi*, male and female. (From Darwin, after Brehm.)

and often woven into its walls also, may be found bright feathers, leaves, colored shells, and various other objects of conspicuous appearance.

After mating comes the preparation for the young. Here again we meet with great variation; the night hawk builds no nest at all, and the plover brings together but a few sticks on which to deposit its eggs; robins, sparrows, warblers, and many other song birds build a more elaborate nest which is usually lined with soft materials which tend to preserve the warmth of the eggs and young. As a rule, primitive birds build crude nests, while birds of a higher type take more care in providing for the safety and comfort of their progeny.

There are a few birds such as the cuckoos and cow birds whose young live at the expense of other species. The eggs of the European cuckoo, for instance, are laid along-

Fig. 153.—Male bluebird with grasshopper. (After Forbush.)

side of the eggs of other birds and the young cuckoo is fed by the rightful owners of the nest like one of their own young. When the young cuckoo develops sufficient strength it has the peculiar instinct of pushing its companions out of the nest, where they frequently perish. Notwithstanding this conduct, which seems like the basest of ingratitude to its benefactors, the birds continue to care for the young interloper until it is ready to take flight.

One of the most pleasing associations with the advent

of spring is the appearance of our feathered friends that are returning from the warmer climes in which they have passed the winter. Scarcely is the snow off the ground, and sometimes even before, when flocks of birds may be seen on their northward journey. Not all birds migrate. Some of the birds of a given locality, like the English sparrows, are *permanent residents*. Some, on the other hand, appear only in winter, having migrated from colder climates; these are the *winter residents*. In most places of the United States a large proportion of the birds pass through the country during their journeys to and from the north; these are the *migrants*. The distances travelled by different species of migratory birds is subject to great variation. Many species (robin, bluebird, meadow lark) winter in the Gulf States or in Mexico, and nest in the northern states or in Canada. Large numbers pass the winter in Cuba and the West Indies, while many species go as far south as the southern part of South America. The golden plover has one of the longest migration routes known. After passing the winter from Patagonia to southern Brazil it does not stop in its northward journey until it reaches its breeding grounds within the Arctic Circle, a distance of nearly ten thousand miles. Migrating birds frequently keep near prominent landmarks, such as coast lines, mountain chains or rivers. The Mississippi valley forms a great highway for hosts of birds, and the same is true only to a less degree of smaller streams. The timbered tracts along the streams form excellent guides for birds flying at any considerable height. Migrating birds commonly fly very high, in some cases at least a mile above the earth, and with their acute vision they are able to survey an immense territory. They have a marvellous ability to find their way back to their old breeding grounds. It is not uncommon for the same birds to nest year after year near the same spot, sometimes in the same nest. In fogs and during stormy nights it is

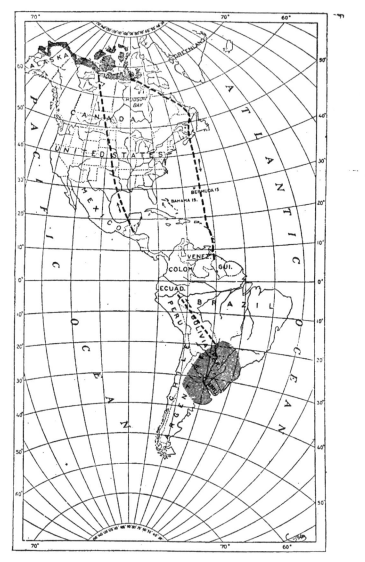

Fig. 154.—Route of migration of the Golden Plover. Breeding range in north finely checkered; winter home in south shown by parallel lines.

true that birds sometimes lose their way; at such times they are especially prone to be attracted to light-houses where many are killed by flying against the walls. In most cases, however, a bird's memory for a route it has once followed is far better than that of the average human being.

The tendency to migrate is doubtless instinctive, but the particular route followed is mainly a matter of experience and tradition. Birds commonly migrate in flocks, and it is probable that it is the older and more experienced birds that lead the way. While it may seem curious that

FIG. 155.—Redwinged blackbird.

birds should leave a genial clime and fly thousands of miles to the cold and barren regions of the Arctic Circle to rear their young, it must be borne in mind that this northern region is one in which during the short summers that occur, there is an abundance of food in the form of berries and insects (especially mosquito larvæ) and comparatively few enemies to molest the young. The migration of birds affords a means of taking advantage of these things. Nature allows few opportunities for making a living to go to waste.

With relatively few exceptions most birds, in one way or another, are of value to man. There are the game birds such as ducks, geese, plovers, snipe, quail and many others whose value is obvious. There are scavengers, such as the sea gulls which devour all sorts of refuse that floats on the water, and the vultures and buzzards which eat de-

FIG. 156.—Yellow-bellied sapsucker.

caying flesh. The accumulated excreta of birds, which is called guano and which occurs in great quantities on certain islands on which the birds congregate, is much employed as a fertilizer of the soil. The plumage of birds is greatly in vogue for purposes of decoration, as well as for various other purposes of a more practical nature. But by far the greatest value of birds lies in their wholesale destruction of insects and other injurious forms of animal life. Much study has been devoted to the food habits of different species birds, especially by the Biological Survey of the United States Department of Agriculture, which has issued numerous bulletins upon the subject. A large part of the data on the food of birds has been accumulated by the examination of the contents of stomachs. By the laborious counting of the different kinds of insects, grains, weed seeds and other bodies which are found in

FIG. 157.—California valley quail. (From Rep. of Calif. Fish and Game Commission.)

the stomachs of many thousands of birds of all sorts a great many very valuable facts have been discovered concerning the utility or harmfulness of various species. Were it not for the destruction of insects by birds it would be difficult for man to raise many of his crops. The common bob-white, or quail, of the eastern and middle states lives mainly on insects and weed seed, and is especially destructive of insects such as grasshoppers, Colorado potato beetles, chinch bugs, army worms, cotton worms, and striped cucumber beetles. Forbush in his valuable book on "Useful Birds and Their Protection" says of the quail that "it is probably the most effective enemy of the Colorado potato beetle." Certain species of birds, such as owls and some kinds of hawks, perform a valuable service in destroying mice, ground squirrels, gophers and other small mammals that are a nuisance to the farmer. Other birds are helpful to the farmer by destroying large quantities of weed seeds. And while

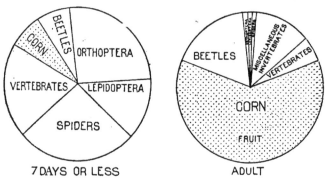

Fig. 158.—Diagram showing proportions of food of American crow (*Corvus americanus*), young and adult. (After Judd.)

many species are destructive to grains and fruits, the damage they do is usually outweighed by the benefit they confer in destroying weed seed and insects. Careful investigation has shown that several kinds of birds commonly deemed injurious are on the whole beneficial. The larger owls which are often shot for their occasional attacks upon poultry are on the whole very valuable birds for reasons above named. There are several hawks which are beneficial since their food consists almost exclusively of small mammals and insects. Others are of more doubtful utility, since they prey upon birds as well as upon small mammals. But there are a few, such as the goshawk, the sharp-shinned hawk and Cooper's hawk, that live mainly upon birds, and hence are an undoubted nuisance.

There are but a few non-predatory birds that are not on the whole valuable to man. A notable exception is the English sparrow which was introduced into this country from England at first unsuccessfully, in 1850 and again with very manifest success in 1853. The species has thrived and multiplied so that there is scarcely a village in the United States that is free from the nuisance. English sparrows eat large quantities of grain while destroying relatively few insects; but their worst offense is their crowding out and destruction of other birds. They demolish the nests of other species, break their eggs,

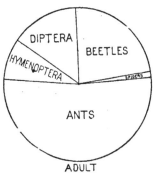

FIG. 159.—Diagram showing proportions of food of barn swallow, *Hirundo erythrogastra*. This bird is almost entirely insectivorous.

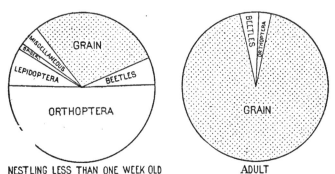

FIG. 160.—Diagram showing proportions of food of English sparrow, *Passer domesticus*, young and adult. (After Judd.)

kill their young and by their continued fighting drive away the older birds. The result is that many more valuable birds have been expelled from towns and villages and considerably limited in numbers in the country. Efforts to exterminate English sparrows have met here and there with success, but to rid the whole country of the pests seems now a hopeless task.

The great value of most kinds of birds has led to the passage of various laws for their protection. For game birds there is usually a "closed season" during which hunters cannot shoot them without incurring a legal

FIG. 161.—A screech owl. (From photo by Mr. Tracy Storer.)

penalty. In several states the song birds and many other kinds are protected at all times. But notwithstanding these protective measures there is a vast slaughter of bird life that is leading to a marked reduction in the number of birds. The robbing of birds' nests by boys destroys many valuable birds, and the still more extensive collection of the eggs of ducks, gulls and other birds which commonly nest upon islands or near rocky shores destroys many more. Aside from the hunting of recognized game birds there is in many places a wholesale destruction for food of such birds as meadow larks, robins, blackbirds, and even many birds of smaller size. Laws prohibiting the sale of game birds prevent the systematic hunting for the market which is so destructive of bird life. Birds are still sacrificed to provide ornaments for ladies' hats, but the combined forces of legislation and public sentiment have diminished the fearful slaughter of our most beautiful birds for this purpose.

One of the most destructive of the enemies of birds is the common house cat. It is unfortunate that this familiar

FIG. 162.—Great grey owl. (From photo by Holliger.)

object of affection and fostering care should prove so mischievous a malefactor, but recent investigations have made the case against pussy a very strong one. Forbush estimates that a mature cat kills on the average more than fifty birds a year and John Burroughs says that cats kill more birds than all other animals combined. Ordinarily the depredations of cats escape notice, since they hunt in a quiet manner and do much of their prowling around at night. Cats are especially destructive to nesting birds and their young. It is true that cats perform a certain service in catching mice and that many individual cats catch few or no birds, but there can be no doubt that, aside from their value as pets, the harm they do in destroying annually millions of birds makes them on the whole a serious nuisance.

Nesting box. (After Forbush).

CHAPTER XXII

THE MAMMALS

The class Mammalia takes its name from the possession of mammary glands which produce milk for the young. All mammals possess these glands and they are found in no other group of animals; consequently their presence serves to define quite precisely this class of vertebrates. Mammals may in most cases be recognized by their covering of hair, just as birds may be distinguished by their feathers. In a few forms, such as the elephant and rhinoceros, the hair is very scarce; and in the whales it has almost entirely disappeared. Porcupines and hedgehogs have many of the hairs modified into large, stiff spines or quills, which are efficient organs of protection. Hair is a product of the outer layer of the skin, the epidermis, and, like feathers, it is commonly shed more or less periodically. New hairs are grown from a papilla at the base of the older hairs that fall out. The nails and hoofs of mammals as well as the outer covering of the horns of cattle and related forms are also epidermal structures.

Unlike birds and reptiles, mammals usually have well-developed external ears, or pinnæ, which are generally shaped so as to catch sound, and are freely movable, as may be seen readily by watching a dog, horse, or rabbit. The sense of smell is generally acute and enables many species to track their prey and others to detect their enemies even at a considerable distance. The least whiff of a human being may send a bear or deer scampering through the forest long before its pursuer appears in sight. By smell mammals may recognize their own kind and distinguish friends from foes within their own species; so the world of smells is one of great importance to them. The keen interest which the dog takes in the various odors encountered along his path is doubtless a trait inherited from his wild ancestors for whom the detection of odors often meant the prevention of starvation or escape from being eaten by some larger animal. It is perhaps because of the keen sense of smell in mammals that there have been developed scent glands in many species which doubtless enable individuals to find one another with greater readiness. Musk which is the product of the scent glands of the musk deer is used as a perfume and also in medicine. In the skunk the scent glands which are situated near the base of the tail are developed to a very unusual degree and constitute an efficient means of defense. The animal can discharge the fluid secreted by the glands to a distance of several feet; consequently there are few creatures that care to molest him, and he can go about with the impudent boldness which is one of the prominent traits of his character.

Mammals are found in most parts of the earth except upon oceanic islands where there are none except bats which may have flown there, or more rarely very small species, such as mice, which may have been transported on floating trees or other drift. Mammals have very limited powers of migration, so that those inhabiting cold countries cannot escape the winter like the birds. Some species have developed a remarkable aptitude for finding food even when the ground is thickly covered with snow; some, such as many kinds of squirrels, store up food during summer which is used during the winter; while others, such as the ground hog, undergo what is called hibernation, remaining in a dormant condition in which they subsist mainly on their own fat.

The demand for many animals on account of their hides and fur has led to a great decrease in their numbers; some, such as the pumas, bears, wolves and wild cats have been rapidly killed off, partly because of their attacks upon domestic animals, partly for the sport of hunting. The species, such as the deer, which supply food are rapidly going. But among the most efficient of the destructive agencies is the wanton killing by hunters for mere sport. For years there has gone on in Africa a fearful slaughter

FIG. 163.—Platypus or duck-bill, *Ornithorhynchus anatinus*. (After Gould.)

of elephants, rhinoceri, antelopes, zebras, giraffes, hippopotimi, gorillas, etc., that has greatly reduced the number of these fine species, so that several of them are threatened with extinction. Many of our finest North American mammals such as the moose, elk, and grizzly bear are comparatively few in numbers and much restricted in range.

The class Mammalia is divided into several orders. The lowest of these, the Monotremes, have the remarkable peculiarity of laying eggs, like the birds and most reptiles, instead of bringing forth living young. The order comprises but three genera, the duck bills (Ornithorynchus), which live in ponds and streams of Australia, the spiny ant eaters (Echidna), which are found in Australia, and a related genus (Proechidna), from New Guinea. The duck bills have peculiar, flattened, protruding jaws like the bill of a duck, and webbed feet. The body is covered with a fine fur. The eggs are laid in burrows. The spiny ant eaters have very narrow jaws; the body is armed with numerous pointed spines amid the hairs, and on the lower

FIG. 164.—Echidna, the spiny ant eater.

side of the female there is a sort of pouch in which the eggs are carried, and then the young, for some time after they are hatched. Aside from the habit of laying eggs, there are various other features which indicate the relationship of the Monotremes to the reptiles; the group may be regarded therefore as in part bridging over the gap between the higher mammals and the reptile-like ancestors from which they were derived.

In the next higher order, the Marsupialia, the young are brought forth alive, but in a very immature condition.

As soon as born they are placed by the mother in a pouch, or marsupium, on the under side of the body. Here they receive nutriment from the mammary glands and are kept warm and protected from enemies. The marsupials are remarkable for their geographical distribution. With the exception of the opossum family they are all confined to Australia and neighboring islands, although remains of their skeletons occur in the deposits of past geological ages in all the continents of the globe. Where the marsupials have been brought into competition with the higher mam-

FIG. 165.—Kangaroo.

mals they have had to succumb; only in Australia which has been isolated from the rest of the world and kept free from serious invasion by other mammals, do we find the marsupials holding their own.

Among the largest of the marsupials are the kangaroos which are remarkable for their long hind legs adapted for jumping, and their short fore legs. There is a great diversity among the marsupials; some are herbivorous, some like the "Tasmanian tigers" are carnivorous, some burrow in the ground like the moles, and still others are arboreal. Nature has adapted them to various modes of life very much as she has done in the case of the higher mammals. The opossum family is represented by several species confined to North and South America. The common Virginia opossum is prized in certain localities for food. Its habit of feigning death, or "playing possum," when captured or cornered doubtless often serves it a good turn by deceiving its enemies. The young after leaving the maternal pouch are carried about on the back of the mother, retaining their hold not only by their claws, but by winding their tails around the tail or limbs of their parents.

Fig. 166.—Virginia opossums. (From Baker.)

The remaining orders of the Mammalia are often termed the *placental* mammals, because the young are retained in the uterus until a comparatively late stage of development, being attached to the uterine wall by a vascular organ, the *placenta*, which serves to convey nourishment and oxygen to the embryo and to carry away its waste matter into the circulation of the mother.

Among the lowest of these orders are the Edentata, which include the sloths, armadillos, and the ordinary ant eaters. The teeth are either absent, as the name of the order implies, or else very poorly developed. The sloths are inhabitants of Central and South America where they live among trees in which they commonly move about along the lower side of the branches, with their backs downward. The armadillos have a peculiar, hard, scaly armor. If threatened with danger the Armadillo rolls up into a ball, like a pill bug, and then becomes quite effectively protected by its horny covering. Most of the armadillos live in South America, but the nine-banded armadillo ranges from Paraguay to southern Texas. The ant eaters are ungainly creatures with long, narrow head and slender extensile tongue. The giant ant eater of South America may reach a length of seven feet, but

FIG. 167.—Nine-banded armadillo.

this is mainly due to its long head and very long, bushy tail. The creatures live upon ants, tearing open ant hills with their strong claws and then throwing out their long, sticky tongues among the disturbed insects which adhere to it and are then drawn into the mouth. As the tongue is worked with great rapidity the ant eater will devour a populous community of ants in a short time.

The odd lot of creatures which constitute the living species of Edentates represent the few survivors of an order of animals once numerous and widely distributed over the earth's surface. One of the best known of the fossil forms is the gigantic Megatherium of South America, which reached a length of eighteen feet. The Glyptodon which sometimes reached a length of twelve feet had a solid rounded armor of bony plates resembling the shell of a huge tortoise.

The order Insectivora includes mostly small animals which (as the name implies) feed mainly upon insects. The best known forms are the moles and shrews. The moles are characterized by their soft fur, small eyes and ears and powerful fore legs with broad hands and strong claws fitted for digging in the earth. Moles spend nearly all their life underground, where they make long burrows, occasionally throwing up mounds of earth, or "mole hills," above their chief habitations. The shrews are small mouse-like animals which lead a very active and mostly underground life in search of insects, snails and earthworms. The European hedgehog has its back covered with a coat of spines, much like the porcupines which belong to a quite different order of mammals, and are often improperly termed hedgehogs in this country.

The order Rodentia, or gnawers, is a very large group of mostly small mammals including the rats, mice, squirrels, gophers, and numerous others. Canine teeth are lacking in the rodents, and the incisors are chisel-like and capable of continuous growth so as to compensate for the wear that results from their frequent use. The rodents are largely vegetable feeders, consequently many species are very destructive. Some of the worst offenders are the common domestic mouse and the house rat, both of which were introduced into this country from Europe. The rat, in addition to its depredations upon grain and all sorts of stored food, is an important agent in the spread of the plague, as has already been described in treating of the flea which may carry this dangerous disease from rats to man. Among the largest of the rodents are the porcupines. The American species ranges over a considerable part of North America. Among the most valuable and interesting of the rodents are the beavers which have been much sought for on account of the value of their fur.

Their webbed, hind feet and broad, flat tail adapt them for their semi-aquatic habits. One of their most remarkable performances is the construction of dams across streams. This is accomplished by cutting down trees which are cut into pieces and dragged into the water where they form the basis about which the industrious animals gather

FIG. 168.—A beaver. (After Baker.)

stones, sticks and various other materials, thus damming up the water. In the water thus backed up the beavers usually construct their mounds, or lodges, which sometimes rise three feet above the surface. These lodges are formed of sticks and stones plastered together with mud and are entered only through the water. In these retreats the beaver passes the winter in security from ordinary enemies.

The group Cheiroptera, or bats, have fore limbs modified into wings. The second, third, fourth and fifth digits of the fore feet are enormously lengthened to form the support of a thin membrane stretched between them and extending from the fifth digit to the hind leg and thence usually to the tail. The first digit, or thumb, is short, and modified into a sort of hook. Our common bats live mainly upon insects which they catch while flying during the night or toward evening. Many of the bats of the old world are fruit eaters, and certain bats of Mexico and

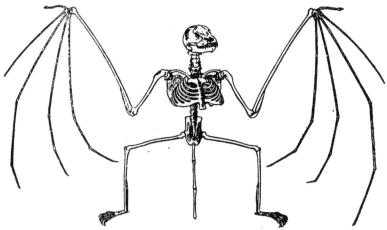

FIG. 169.—Skeleton of a bat.

South America live by sucking the blood of mammals. They attack their victims while the latter are asleep; and, after making incisions with their very sharp teeth, quietly suck the blood. Many bats, but by no means all, have very small eyes. While the popular notion that bats are blind is incorrect, these animals when blinded may fly about a room full of objects or even strung with wires without encountering any of the obstructions. They apparently have other very acute senses which warn them of the proximity of obstacles to their flight. Bats gener-

THE MAMMALS 219

Fig. 170.—A zebra. (After Baker.)

Fig. 171.—Hippopotamus. (After Baker.)

ally hibernate in caves, hollow trees or other dark places to which they commonly retreat during the day.

The order Ungulata, the hoofed animals, constitutes one of the largest and most important groups of mammals. In general ungulates are herbivorous; their molar teeth are fitted for grinding and the canines are usually small or wanting. The order is commonly divided into the

FIG. 172.—African elephant. (After Baker.)

odd-toed ungulates (Perissodactyla), such as the horse, tapir and rhinoceros; the even-toed ungulates (Artiodactyla), such as the ox, sheep, deer, camel, pig, hippopotamus; and the Proboscidea, or elephants.

Some of the even-toed ungulates such as the deer, sheep and oxen are called *ruminants* on account of their habit of chewing a cud. Most of these animals have a four-chambered stomach, one division of which receives the food when first swallowed, and later regurgitates it into the mouth where it is thoroughly chewed at leisure. Watch a herd of cattle when they are standing quietly after feeding and you will probably see them engaged in the satisfying operation of ruminating. We use the term ruminate in a figurative sense signifying to ponder quietly over any subject, but though cattle look as if

FIG. 173.—American bison and calf. (After Allen.)

they might be engaged in solemn reflective thought it is reasonably certain that the appearance is quite deceptive. Most ruminants are provided with horns which may be either hollow and fitted over a permanent bony core, as in the ox family, or Bovidæ; or solid, as in the deer family, or Cervidæ. The solid horns, or antlers, are generally shed annually, breaking off at the base. While they are being renewed the horns are covered with a furry skin, or "velvet," which becomes dead or dried when the horns are full grown when it becomes peeled off. Antlers are generally found only in the male, but the reindeer has them in both sexes. Their usual limitation to the male sex may be correlated with the fact that the males employ them frequently in fighting with one another, especially during the breeding season. In the

FIG. 174.—California valley elk. (From a group in the museum of the California Academy of Sciences.)

United States, deer and the larger related forms, the moose and elk, have rapidly diminished in numbers, and game laws are enacted for their protection from hunters.

Related to the deer family are the antelopes which are represented in North America by the beautiful pronghorn antelope of our western plains. Of the hollow-horned ruminants the once abundant but now nearly extinct buffalo and the Rocky Mountain sheep are the best known of our native species. None of the odd-toed ungulates are native to the United States. The wild horses and asses of both North and South America are descendants of animals brought over from the old world.

The order Carnivora includes the flesh eaters, viz., tigers, lions, wolves, dogs, cats, bears and many other related

FIG. 175.—The American prong horn antelope. (From a group in the museum of the California Academy of Sciences.)

forms. They have soft, padded feet and four or five well-developed toes, with strong claws which in the cat family are curved and retractile and adapted for seizing prey. The teeth are fitted for tearing flesh rather than for grinding, and there are usually well-developed, pointed canines. The cat family, Felidæ, includes some of the largest of the flesh eaters, such as the lions, tigers and leopards of the old world, and the jaguars and pumas of the new. The puma formerly ranged over a larger part of temperate North America, but in the United States it is now limited to the west. Formerly it inspired much fear in the early settlers who commonly called it the panther and related many thrilling stories of its ferocity. As a rule, however, it is exceedingly shy of man and is only rarely seen. The somewhat larger South American jaguar ranges north through Central America and Mexico into southern Texas. While the cat family is exclusively carnivorous, the bears and their smaller allies, the raccoons, will eat many other

Fig. 176.—African leopard.

Fig. 177.—A raccoon. (From Baker.)

things besides flesh, such as nuts, berries, acorns and the leaves of plants. The largest species of bear in the United States, the grizzly, is now found only in a few remote localities in the west.

The animals of one division of the carnivora, the Pinnipedia, have taken to an aquatic life. These are the seals, sea-lions and walruses. While resembling other carnivores in fundamental features of structure, the pinnipeds have undergone striking changes of external form in adaptation to living in the water. The fore legs are modified into fin-like flippers fitted for swimming, and the hind limbs are flattened and directed backward. Most of the pinnipeds are inhabitants of the colder parts of the ocean. The seals especially are much sought after for their fur and they have consequently decreased considerably in numbers, so that it has been found necessary to protect their breeding places by law.

The most exclusively aquatic of the mammals are the Cetacea which include the whales, dolphins, porpoises

FIG. 178.—Habitat group of Steller's sea lions showing large male, females and young. (From a group in the museum of the California Academy of Sciences.)

and their allies. The whales are by far the largest of the mammals, the largest whale, the sulphur bottom of the Pacific Ocean, reaching a length of ninety-five feet and a weight of two hundred and ninety-four thousand pounds. The general form of a whale is more or less like that of a fish; the fore legs are modified into flattened flippers and the tail is expanded, but it differs from the tail of a fish in being flattened horizontally instead of vertically. The hind limbs of whales have almost disappeared, being represented by **minute rudiments** buried deep within the flesh. In the whalebone whales the teeth have entirely disappeared in the adult, but they appear during the early development of the embryo, thus indicating the descent of these whales from toothed ancestors. What is commonly called whalebone is not really bone, but a horny substance that occurs in the form of a fringe of plates attached to the upper jaws. This fringe serves as a strainer to hold in the creatures the whale catches in its capacious mouth while allowing the water to pass through. Whales feed upon fish, squid and various small animals that swim in the open ocean. Hairs in the whales are almost entirely absent. Heat is retained in the body by means of the very thick layer of fat, or blubber, beneath the skin. It is this fat that yields whale oil, the pursuit of which led to an extensive whaling industry which was carried on until the supply of whales became greatly reduced. Spermaceti is a product of an oil which comes from a large cavity in the head of the sperm whale. The latter differs from the whalebone whales in having numerous conical teeth instead of plates of whalebone. The nostrils of whales are united to form a single aperture on the upper surface of the head. As air is blown out of this opening, from the lungs, a column of spray, the condensed moisture of the breath, appears which has given rise to the erroneous notion that the whale spouts out the water taken in through the mouth. Whales may remain under water for a long time, but like all animals that depend upon their lungs for their supply of air they are compelled to come to the surface to breathe.

The sea cows constitute a small order of aquatic mammals, called the Sirenia. These animals live in rivers or near the shore of the ocean where they feed upon aquatic vegetation. One of the largest species, Steller's sea cow, was an animal of twenty to thirty feet in length, inhabiting the Behring sea. As the animals showed little fear of man they were entirely exterminated in the eighteenth century by hunters who killed them for food. Another species occurs on the coast of Florida and a few others in the old world.

The Primates which comprise the lemurs, monkeys, apes and man constitute the highest of the mammals. The higher primates approach man in the general form of the body, the occasional upright position, opposable thumb which permits the foot to be used as a hand, and in many other features of structure. The lower primates, the lemurs, generally go on all fours and have a protruding muzzle like that of a dog. The lemurs are confined to Africa, the Orient and Madagascar, being most numerous in the latter island where they constitute a large part of the mammalian fauna.

The primates of the new world have a broad nose and usually a long tail which is employed to wrap around the branches of trees from which the monkeys frequently suspend themselves. There are many species of rather small size confined to South and Central America, but none occur native in the United States.

The old world primates are much more varied in character. While they include many interesting forms, the chief interest attaches to the large, anthropoid (man-like) apes which are of all animals the most closely related to man. The anthropoids include the gorillas, chimpanzees, orangutans and the gibbons, the latter being the lowest and most monkey-like of the group. The orangs are confined to Borneo and Sumatra where they live mainly in trees. The chimpanzees are natives of Africa living mostly in wooded regions. While commonly walking on all fours chimpanzees are capable of walking erect and they use their hands in a very human way. They are known to throw

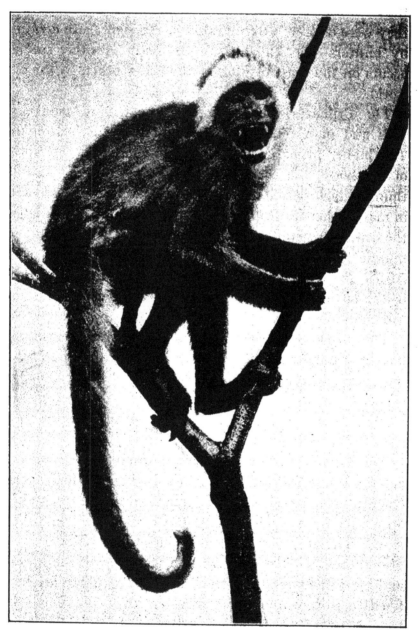

Fig. 179.—A South American arboreal monkey, Cebus. (From photo by Sanborn, with permission of the New York Geological Society.)

sticks and stones at their enemies, and to use sticks for pulling in objects otherwise out of reach. Trained chimpanzees have been taught to ride a bicycle, skate on roller skates,

Fig. 180.—Young chimpanzee.

eat at a table with knife, fork and spoon, and to select one of a bunch of keys and unlock a lock. Although savage fighters, chimpanzees are very fond of their mates and offspring, and often manifest intense grief over the loss of one of their family group. The father and mother of a family appear to remain together permanently. Either sex may carry the young offspring about in the arms much as human beings carry their babies.

The largest and strongest of the apes is the gorilla, a native of Africa. The gorilla lives in forests, often constructing a sort of nest in the branches of a tree where it spends the night. Powerful, brutal, ferocious, the gorilla is a formidable foe; but one which rarely if ever attacks man unprovoked. Gorillas have never been kept very long in captivity. They sulk, lose their spirit, and if they do not succumb to tuberculosis, which kills so many of the apes in captivity, they languish and die, apparently through sheer mental depression.

The older naturalists set man apart in a group by himself, but as Huxley conclusively showed by a very thorough comparison of the structure of man with that of the apes, man differs less in structure from the higher apes than the latter differ from the lower members of the order of primates. Man differs greatly from the apes in the size of the brain, which is a certain though inadequate index of his greatly superior intelligence. There is abundant evidence that the apes are by far the most intelligent of the animals below man; but it is quite certain that the intellect of the lowest savage stands very high above that of his highest simian relative. There are no "connecting links" between man and ape at present living on the earth, but in ancient deposits fossil bones of human beings have been discovered which probably belonged to a type of man more primitive than any existing race.

Man is regarded as constituting a separate family, the Hominidæ, and a single genus, Homo. The various kinds of men, notwithstanding their marked differences, are commonly considered as members of a single species,

Homo sapiens. Man's permanently upright walk, hairless body, and large brain are among the more conspicuous differences in structure separating him from the apes, but there is nevertheless an astonishing similarity in many other structural peculiarities as well as in embryological development. In a later chapter some of these characteristics will be briefly described.

CHAPTER XXIII

THE CHEMICAL BASIS OF LIFE

The division of Biology which deals with the activities or functions of the parts of a living organism is known as *physiology*, as distinguished from *morphology* which deals with the structure of organisms. The activities of an organism are in part physical and in part chemical. The material world both living and non-living is made up of a limited number (about eighty) of substances, called *elements* which cannot be separated further into substances of different kinds. Different elements may be grouped together to form *compounds* and compounds may be separated into their elements. Thus common salt is a compound composed of two elements sodium and chlorine, but the elements themselves cannot ordinarily be decomposed further. Changes which involve the combination, separation or rearrangement of elements are known as *chemical* changes. The rusting of iron is a chemical change. Iron, an element, combining with the oxygen of the air, another element, produces a substance, iron oxide, which is very different in appearance and properties from either of its constituents. Other chemical changes are the burning of wood and coal, the action of acid on soda and the fermentation of sugar. In all these cases there are changes between the elements of the substances involved in the process, resulting in the production of very different kinds of substances. Changes which do not involve any alteration of the elementary composition of bodies are called *physical*. The conversion of water into steam or ice, the solution of sugar in water, the mechanical movement of bodies, and the propagation of sound, light and electricity are physical changes and their treatment belongs to the science of *physics*.

One striking peculiarity of chemical changes is that they occur between certain definite proportions of the elements involved. This principle is called "the law of definite proportions." Thus sodium and chlorine always combine in a certain definite ratio by weight, 23 parts of sodium to 35.5 parts of chlorine to form common salt, or sodium chloride. If more of one or the other element is present it simply remains uncombined. Sometimes different elements may combine with each other in more than one ratio, but the different ratios have a simple relation to one another. Thus carbon and oxygen may combine in the ratio of twelve parts of carbon to sixteen parts of oxygen to form a gas, carbon monoxide (CO), and also in the ratio of twelve parts of carbon to thirty-two of oxygen to form carbon dioxide or carbonic acid gas (CO_2) which has properties very different from those of carbon monoxide. In carbon dioxide there is just twice as much oxygen in the compound as in carbon monoxide. The definite numerical ratios in which elements unite into compounds form one of the several considerations that have led men of science to the view that chemical elements are made up of minute, indivisible bodies called *atoms*. The atoms may be united into groups called *molecules*, a molecule being the smallest part into which a compound may be divided without losing its properties. The division of a substance into molecules may involve nothing but physical changes, but to divide a molecule into its constituent atoms would constitute a chemical change.

In order to express the chemical constitution of bodies

in a convenient form chemists have given the elements certain symbols, H for hydrogen, O for oxygen, C for carbon, N for nitrogen, S for sulphur, K for potassium (kalium), Na for sodium (natrium), etc. The chemical composition of bodies may be indicated by groups of symbols representing their constituent elements: salt, NaCl; water H_2O; carbon dioxide, CO_2; sulphuric acid, H_2SO_4; etc. Each molecule of common salt is supposed to contain one atom of sodium and one atom of chlorine; each molecule of water two atoms of hydrogen to one of oxygen; and each molecule of sulphuric acid, two atoms of hydrogen, one of sulphur and four of oxygen.

Chemical changes or reactions are expressed in the form of an equation; thus the formation of water by the addition of oxygen and hydrogen is indicated by $2H + O = H_2O$ and the decomposition of calcium carbonate by heat,

$$\underset{\text{Calcium carbonate}}{CaCO_3} = \underset{\text{Calcium oxide}}{CaO} + \underset{\text{Carbon dioxide}}{CO_2}$$

Besides the chemical changes resulting from the simple combination or dissociation of elements as in the two illustrations just given, we may have, in bringing together two compounds, an exchange of certain of their elements. Thus putting sodium chloride, NaCl, and sulphuric acid together, the sodium and the hydrogen of the two compounds exchange places.

$$\underset{\text{Sodium chloride}}{2NaCl} + \underset{\text{Sulphuric acid}}{H_2SO_4} = \underset{\text{Sodium sulphate}}{Na_2SO_4} + \underset{\text{Hydrochloric acid}}{2HCl}$$

Living matter is composed of comparatively few elements. Certain of these are found in all organisms while the occurrence of others is less widespread. Descriptions are here given only of the more important ones.

Oxygen

Oxygen is a transparent, odorless gas which unites readily with a large number of elements and compounds. It is one of the most abundant elements, as it occurs in water, and forms, in combination with various minerals, about one-half of the earth's crust. Air is composed of about one part of oxygen to four of nitrogen together with small quantities of water vapor and other gases. The oxygen of the air is simply mixed with nitrogen and not chemically combined with it. Oxygen in a pure state may be obtained by the decomposition of water by the electric current or by heating various substances that contain it in chemical combination. In its pure state it acts very vigorously upon many substances that it attacks but feebly in the air. Thus a steel watch spring will burn in pure oxygen and a glowing match thrust into pure oxygen will quickly burst into flame. Most of what is called combustion or burning is the combination of substances with oxygen. When wood and coal are burned they combine with the oxygen of the air giving rise mainly to carbon dioxide and water. The combination of substances with oxygen is called *oxidation*, a process which may be a rapid chemical change such as takes place in the explosion that occurs when oxygen and hydrogen are mixed and ignited, or a very slow one such as the gradual rusting of iron. Oxidation plays an essential rôle in the living body. Oxygen occurs in all living tissues, and it is found in all foods.

Carbon

Carbon is a solid devoid of taste or odor. It may be seen in almost pure form in charcoal. When burned it combines with oxygen to form a gas, carbon dioxide.

This gas is a common product of living bodies, since it results from the action of oxygen on the carbon contained in living tissue. Carbon occurs in all tissues and in all foods.

Hydrogen

This element naturally occurs in all organisms since it is one of the constituents of water, but it is found also in other combinations in all living substance. It is a very light, transparent, odorless gas that enters very freely into composition with oxygen, chlorine, carbon and a number of other elements.

Nitrogen

This element is a rather inert gas, transparent and odorless, as we might infer from its constituting about four-fifths of the atmosphere.

Other elements contained in living bodies are sodium, potassium, calcium, sulphur, phosphorus, chlorine, iron, iodine, and in some cases silicon, manganese and copper. Many of these elements occur in the form of salts which, while not commonly classed as foods, are nevertheless necessary to maintain the life of the body. The elements of the living body are for the most part combined to form substances of a good deal of complexity. Most of the compounds formed are not found elsewhere in nature, and they are consequently known as organic compounds. It was formerly held that organic compounds could be formed only through the agency of life, but chemists have succeeded in making a good many of them artificially in the laboratory. The very complex and unstable compounds more immediately associated with the phenomena of life it is still impossible to fabricate. The body of the simplest organism is a chemical laboratory in which processes go on that are far more complex than those which the chemist has been able to control.

Animals differ from most plants in requiring organic compounds for their food. Most plants are able to manufacture their living substance from the inorganic constituents of the air and soil, but animals are compelled to live upon plants or other animals which furnish food in the form of organic matter. Organic food substances fall into three principal classes: *proteins, carbohydrates* and *fats*. Proteins are complex compounds containing C, H, O, N, and frequently other elements. The white of egg, cheese, and the lean fiber of meat consist almost entirely of protein. All living matter contains protein material which alone can supply the nitrogen for the animal body. Fats are more or less oily substances containing C, O, and H. They are quite readily oxidized and yield a considerable quantity of heat. Butter, olive oil, suet, lard and tallow are common examples of fatty substances.

Carbohydrates are composed of C, H and O, there being twice as many atoms of H as O in the molecule. They include such substances as sugar, starch and cellulose. Starch is commonly stored in the cells of plants in the form of grains with concentric layers like the coats of an onion. When treated with iodine it turns blue. It is insoluble in water, but it may be converted into sugar by fermentation.

This process of fermentation is one of the most common kinds of chemical action that takes place in organisms. It may be illustrated in the fermentation of sugar or molasses. If a small amount of yeast is added to a solution of sugar, after a time small bubbles of gas (CO_2) may be seen to arise from it, and its temperature increases. After the process has run its course the sugar in the solution disappears and in its place there is found a certain amount of alcohol. The conversion of grape juice into wine depends upon the fermentation of the sugar of the fruit into alcohol and carbon dioxide. The fermenting agent is here the yeast plants, minute plant organisms which rapidly multiply in the liquid. Alcohol itself may be fermented by other organisms that convert it into acetic acid which is the acid of vinegar.

There are a great many kinds of fermentation caused by different species of yeast and bacteria. Fermentation may be caused also by certain substances called *ferments* or *enzymes*. Ferments are produced by organisms and they have the remarkable property of converting a great many times their own bulk of other substances. A very small amount of an enzyme may convert a very large amount of a material without suffering any appreciable loss. Human saliva contains a ferment, *ptyalin*, that converts starch into sugar. In the gastric juice, which is a secretion of the stomach, there is another ferment, *pepsin*, which converts proteins into simpler compounds called peptones. By fermentation complex substances are split up into simpler ones and these again may be further split up by other ferments. Heat is liberated during fermentation and energy is thus supplied to the body. Under certain conditions enzymes may build up more complex compounds out of simpler ones, thus affording a means of keeping up the supply of complex substances in the body.

The real living substance of an organism, or in Huxley's phrase "the physical basis of life," is commonly called *protoplasm*. This is a semifluid substance of a great chemical complexity, and it differs somewhat in composition in different species of animals and plants. It contains, C, H, O, N, and frequently S, P and K. Unlike the substances just described, protoplasm has the power of active growth, taking up food materials of various kinds and converting them into its own substance. This process which is known as *assimilation* is an essential attribute of all living material. The assimilated material is not added to the outside as in the growth of stones and most crystals, but permeates the entire mass. An animal may live upon various other kinds of protoplasm, but the foreign protoplasm is broken down and absorbed, and then worked over in the wonderful chemical laboratory of the living tissues into the peculiar protoplasm of the devouring animal.

Along with the assimilation of food, protoplasm is continually undergoing a process of breaking down or waste, and the materials so formed are got rid of. This process is called *excretion*. A living organism may thus be compared to a vortex through which matter is continually passing; the food taken in is broken down and built up into living substance which after a time is broken down again and eliminated. The form of the organism, like that of a vortex or a waterfall, may remain constant, but the matter of which it is composed is subject to a continual change.

All protoplasm requires oxygen. The oxidation of protoplasm supplies heat and other forms of energy just as the oxidation or burning of coal in a furnace supplies heat for the running of an engine. Among the most common products of oxidation in living matter are carbon dioxide and water which are the same compounds that result from the burning of a candle or a piece of wood. The union of organic substance with oxygen and the giving off of the products of this union (CO_2 and H_2O) is called *respiration*. The cessation of respiration results in the death of an organism just as the withdrawal of oxygen will quickly put out a fire.

CHAPTER XXIV

CELLS AND TISSUES

One of the most remarkable of the properties of "the physical basis of life" is its tendency to build itself up into a definite form and structure. Cattle, sheep, insect larvæ, and even parasite fungi and bacteria may all live upon the tissue of the same kind of plant, each organism converting the plant's tissue into its own peculiar kind of protoplasm. We ourselves eat many kinds of meat and vegetables, but these foods are all converted into our own living substance. The form assumed by each creature depends very little upon the food it assimilates, but very much upon the chemical and physical properties of the protoplasm that is peculiar to its species. Each kind of protoplasm tends to produce its own particular kind of organization, be it man, dog, worm or plant.

The body of a higher animal such as man is a wonderfully complex mechanism, and in order to carry on its many kinds of activities it is divided up into different parts or *organs* each of which is especially fitted for its peculiar work. Thus we have organs of locomotion, organs such as the stomach, liver, etc., for the work of digestion, organs of circulation for propelling and conducting the blood, organs of respiration, excretion, and many others. In a very simple animal such as the Amœba these various activities or functions are carried on by all parts of the jelly-like body. There are no especial organs for the function of respiration or digestion or any other activities. Food is taken in anywhere and digested anywhere in the interior of the animal. Respiration takes place all over its surface and there is a constant circulation of the living substances.

As we pass up the scale to an animal such as the freshwater Hydra we find digestion carried on in the interior cavity of the body; there are definite organs, the tentacles, for the capture of food; there are muscle fibers which by their contraction change the form of the body; and there are special nettling cells set apart for the purpose of protection. Nevertheless respiration is carried on, not by special organs, but over the entire surface of the body; all parts of the body apparently eliminate waste products; there is no blood, nor are there organs of circulation. Special organs have been set apart for some kinds of work, while other functions are discharged by the body in general much as in the Amœba.

When we come to higher animals there are organs especially fitted for respiration, such as the gills of fishes and the lungs of mammals; special organs are adapted for excretion, such as the kidneys; and other organs are exclusively concerned with the circulation of the blood. Passing up the scale of life more and more organs are added; each becomes especially fitted for its work and at the same time less able to do the work of other parts.

The organs of a higher animal are formed of different kinds of materials called *tissues*, and most tissues are in turn composed mainly of small bodies called *cells* which bear somewhat the same relation to the organism as a whole as bricks bear to a brick house. Cells are masses of protoplasm, commonly, though not always, surrounded by a membrane or cell wall, and containing a small vesicle known as the *nucleus*. The latter structure is an essential element in the life of the cell. Cells multiply by a process of fission, the nucleus dividing along with the protoplasm of the cell body. In a certain sense, cells may be regarded as little organisms, capable of growth, and multiplication; the body being, as it were, a society of cells. As cells have such varied things to do in the work of the body it is very natural that they have different shapes and structures, and the several varieties of tissues owe their differences largely to the different kinds of cells of which they are composed. The more common tissues fall into the following classes:

Epithelium.—Epithelial tissues occur commonly in the form of layers, the cells of which fit very closely together.

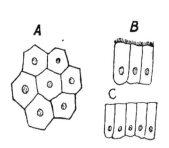

FIG. 181.—Forms of epithelium. A, flattened or squamous; B, ciliated; C, columnar.

FIG. 182.—Fibrous connective tissue showing fibers and a few cells.

Such tissue is found covering the outer surface of the body and lining its various cavities such as the alimentary canal, the cœlom, or body cavity, the interior of blood vessels, etc. Sometimes the cells are very thin and flattened (*squamous epithelium*), or sometimes nearly cubical (*cuboid epithelium*), and very often elongated (*columnar epithelium*). In the variety known as *ciliated epithleium* the free edges of the cells are covered with cilia or short hair-like processes which beat to and fro creating a current in the liquid bathing the cell.

Connective tissue is composed usually of scattered cells

between which occurs more or less intercellular substance. The general function of connective tissue is to hold various parts together and to act as a supporting substance. It occurs in bone and in cartilage (gristle), in ligaments and tendons, and in the form of membranes and networks binding together the cells of various organs.

 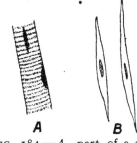

Fig. 183.—Cartilage. Fig. 184.—*A*, part of a striated muscle fiber; *B*, fibers of unstriated or involuntary muscle.

Muscular tissue is composed of elongated cells which have the property of contracting strongly when they are stimulated. In striated muscle which composes the great

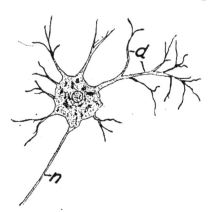

Fig. 185.—Nerve cell. *d*, branching dendrites; *n*, process forming the axis cylinder of a nerve fiber.

mass of the muscles of our limbs and body the cells are marked with longitudinal striations due to the pressure of very fine thread-like structures called *fibrillæ*. There are also cross striations running across the cell, and generally several nuclei in each fiber. In unstriated muscle the cells are generally smaller, and have no cross striations and there is one nucleus to a cell.

Nervous tissue is composed of cells and fibers which make up the nervous system.

Most organs of the body are composed of several tissues. The heart for instance consists mainly of muscle fibers, but these fibers are held together by connective tissue. There is a coat of epithelium lining the cavity of the heart as well as its outer surface, and there are nerve cells and fibers embedded in the heart muscle.

CHAPTER XXV

DIGESTION

What we call living embraces a multitude of different activities. In your bodies while you are reading this paragraph, muscles are contracting, nerves are conducting stimuli, air is being drawn into and expelled from the lungs, the blood is surging through the blood vessels, absorption is taking place through epithelial membranes, waste matter is being eliminated from the blood, every cell is being built up and torn down, chemical changes are tak'ng place in every bit of living matter. It is these chemical changes that keep your body warm and supply the energy for its various activities. Were the chemical changes to stop, everything else would stop, and the body would become inert and cold. We might compare our bodies to a steam engine whose supply of energy comes from the burning of coal in the furnace, or in other words the chemical union of carbon and perhaps a certain amount of hydrogen with the oxygen of the air. The movement of levers and wheels depends on the expansion of steam which is caused by the heat generated by chemical changes. Without the burning coal the engine would be inert and cold. To keep the engine running, more coal is continually added to the furnace, the ashes or unburned waste are removed, and there are arrangements for removing the smoke and carbon dioxide resulting from the burning of the fuel.

What fuel is to the engine food is to our bodies. Our food supplies not only the material from which our bodies are built, but it furnishes the energy for performing the work of the body. In order to yield this energy the food must undergo chemical decomposition. It is split up by ferments, and then oxidized or burned in the various tissues. Some of these chemical changes result in the building up of living matter out of simpler substances. Others result in its tearing down or decomposition.

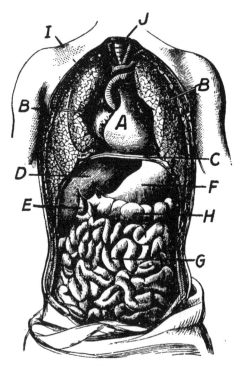

FIG. 186.—*A*, the heart; *B*, the lungs drawn aside to show the internal organs; *C*, diaphragm; *D*, liver; *F*, gall cyst; *E*, stomach; *G*, small intestine; *H*, large intestine. (After Le Pileur.)

In vital activity we therefore have a twofold process of waste and repair—a process known as *metabolism*—which goes on continually. Metabolism is the very care and essence of vital activity; all other processes are dependent upon it. But as it always involves waste, it always requires new material or food to enable it to keep going.

In order that the various organs of the body can receive food material for their growth and activities the food must be prepared for being assimilated. Such preparation is the work of the organs of digestion. These organs act upon the food so as to convert it into a soluble form capable of passing by osmosis into the tissues where

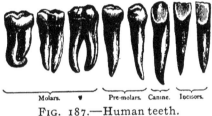

FIG. 187.—Human teeth.

it is assimilated. The conversion of food into soluble form commonly involves both mechanical and chemical processes. In man and many other animals food is chewed so that it is divided up into smaller particles which can be acted on more readily by the digestive juices. Another mechanical process consists in moving the food through the alimentary canal where it is acted upon by ferments and absorbed. These functions involve various organs which will now be described.

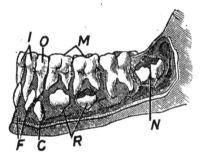

FIG. 188.—Teeth of child of six and one half years. *I*, incisors; *O*, canine; *M*, molars; *E*, permanent incisors; *C*, canine; *R*, bicuspids; *N*, molar. (After Marshall.)

The Teeth.—In man there are thirty-two teeth. The two front pairs in each jaw are the cutting or *incisor* teeth. Just outside of these are the *canine* or *eye* teeth; then come two *bicuspids* on each side, and finally the three *molars* or grinding teeth. The last of the molars appear quite late in life and are commonly called the *wisdom teeth*. The first set of teeth, or milk teeth, which the child begins to get usually in his first year, are but twenty in number. These are later replaced by the permanent teeth. Each tooth is set into the jaw bone by one or more roots. There is a cavity in each tooth filled with *pulp* which contains blood vessels and nerves and is sensitive to pain. The body of the tooth is composed of a bony substance called *dentine*, and the exposed part is covered by a very hard layer, the *enamel*.

In animals below man the teeth vary much in number, size and shape. In the carnivores they are fitted for tearing flesh, while in the herbivores they are adapted for grinding food. The incisors are well developed in the rodents or gnawers, while in the elephant the upper incisors are modified into the enormous tusks which afford us our ivory. In fishes, amphibians and reptiles the teeth are generally conical and fitted for seizing prey, which is their primitive function. These animals as a rule do not masticate their food but swallow it whole. Teeth are subject to decay which is usually caused by the lodgment of particles of food that undergo decomposition. When the decay reaches the pulp cavity we are generally reminded of the fact and are compelled to seek the services of the dentist. Keeping the teeth clean by frequent use of the tooth brush prevents the beginning of trouble.

The Salivary Glands.—Three pairs of salivary glands pour their secretion, the saliva, into the mouth cavity. The saliva is an alkaline fluid containing a considerable amount of mucus and a ferment, *ptyalin*, which converts starch into sugar. Chewing food not only divides it into smaller particles, but it mixes it with saliva which thus has a better opportunity to act upon the starchy constituents.

The Stomach and Gastric Digestion.—When we swallow our food it passes through a long tube, the *esophagus*, into the stomach. The stomach is a muscular organ which tapers toward the end away from the esophagus where it is furnished with a muscular constriction, the *pylorus*. The pylorus when contracted serves to retain food in the stomach until it is digested. The stomach is lined with a mucous membrane filled with numerous gastric glands. These glands secrete the gastric juice, which is slightly acid from the presence of a small amount of hydrochloric acid (about 0.2 per cent.). They also produce a ferment called *pepsin* which acts upon the protein substances of our food, converting them into a soluble form called *peptone*. Pepsin has no action upon fats or carbohydrates, and it acts upon proteins only in an acid medium. The presence of food in the stomach excites the secretion of the gastric juice, and the contractions of the muscular walls of the stomach which are set up by the same cause produce a sort

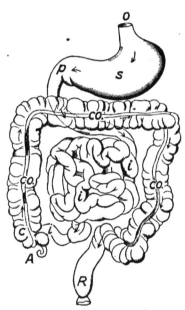

FIG. 189.—Diagram of stomach and intestines. O, esophagus; S, stomach; p, pylorus; i, small intestine; co, colon or main part of large intestine; R, rectum or terminal division of large intestine; A, appendix vermiformis attached to the cecum, c. The arrows indicate the directions taken by the food.

of churning motion which mixes the gastric juice with the food and indirectly aids in the process of digestion. When the food has been acted on for a time in the stomach, the pylorus relaxes and allows the more or less fluid mass to escape into the small intestine. The latter is a long, coiled tube with rather thin, muscular walls and an inner lining of mucous membrane which contains numerous glands. It is the contraction of the muscular walls that forces along the food. The inner surface of the small intestine is thrown into many folds, and these are beset with numerous minute

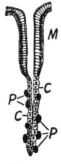

Fig. 190.

Fig. 191.

Fig. 190.—Gland of stomach. *M*, mucus forming cells; *C*, chief cells; *P*, parietal cells.

Fig. 191.—Part of small intestine cut open to show folds in lining.

finger-like projections called *villi* whose function it is to increase the surface available for absorption. The villi are richly supplied with blood and lymph vessels which carry away the soluble food materials absorbed from the intestines.

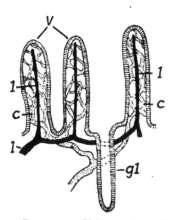

Fig. 192.—Villi, *v*, of small intestine; *c*, capillaries of blood vessels; *gl*, gland; *l*, lacteals or lymph vessels.

The small intestine leads to the large intestine. Near the point where the latter begins there is a short pocket, the *cæcum*, at the end of which is a small tubular organ, the *vermiform appendix*. Whatever may be the function of the latter organ it is often the seat of inflammation (appendicitis) which frequently necessitates an operation for the removal of the offending part.

Two large and important glands, the *pancreas* and the *liver*, pour their secretions by a short common duct into the small intestine. The pancreas secretes an alkaline fluid, the pancreatic juice, which contains three ferments; one, *amylopsin*, which converts starch into dextrose, a form of sugar; another, *trypsin*, which digests proteins; and a third, *lipase*, which acts upon fats transforming them in part into soap and converting them into an emulsion. The materials which escaped digestion in the stomach are acted upon by the pancreatic juice in the small intestine.

The liver is a very large, reddish organ lying near the stomach in the right side of the abdomen. Its secretion, the *bile* or *gall*, may accumulate in a small sac called the *gall bladder* before it is discharged into the intestine. The bile consists in part of waste products, and it plays little part in digestion, but it facilitates the absorption of food through the walls of the intestine. Besides the secretion of bile the liver performs another important function in storing up a sort of reserve food supply in the form of *glycogen*. This substance is a carbohydrate, allied to starch. The carbohydrates which are absorbed by the blood in the form of sugar (dextrose) are in part converted by the liver into glycogen which accumulates in the liver cells. At other times, especially during hunger or severe exercise, glycogen is converted into sugar which is given off into the blood. The liver, therefore, serves as a sort of temporary storehouse, converting the excess of carbohydrate into the comparatively insoluble form of glycogen which is given out again in times of greater need.

The small intestine, which is the seat of important processes of digestion, is an organ especially adapted for the absorption of digested food. A certain amount of absorption occurs in the stomach, and also in the large intestine, but most of it occurs in the small intestine whose numerous villi and folds with their rich supply of blood and lymph vessels afford a large surface through which the soluble products of digestion have ready access to the blood and lymph.

CHAPTER XXVI

FOODS AND THEIR USES

After the food is digested and absorbed it is carried by the blood and lymph to various parts of the body where the different organs convert it into their own peculiar substance. This conversion is the process of *assimilation*, to which digestion and absorption were merely preparatory. It is one of the most wonderful as well as one of the least understood of the activities which take place in the living organism. Assimilation not only compensates for the waste that is always being produced by the body, but it enables the body to increase in size. When waste is exceeded by repair as in the normal small boy there is growth. In a fever when the tissues are rapidly consumed, or burned, there is loss of weight; this may be very marked if the fever is severe. Growth is rapid in the early years of life and a relatively large amount of food is required as is evinced by the keen and frequently recurring appetite of healthy youth.

The many articles of diet which we consume are quite different in their chemical composition and they are put to different uses in the economy of the body. The true living substance, or protoplasm, requires for its formation foods which contain all the necessary chemical elements. Since proteins contain nitrogen (this element is absent in carbohydrates and commonly in fats) a certain amount of protein is absolutely necessary for the continued maintenance of life. Fat in the body may be derived from carbohydrates or from other fats. Many people find sugar and starchy foods fattening. But no amount of starch or fat would prevent a man from dying of starvation, because he requires some food which contains nitrogen.

The carbohydrates and fats, while they do not alone suffice to form the living protoplasm of our body, nevertheless supply us with the energy that maintains our bodily heat and enables us to do muscular work. Proteins also supply us with energy in addition to affording all the elements necessary for building up living tissue. We might live on protein food alone, but in order to supply the energy we need we should have to eat so large a quantity of protein that there would be an abnormally large amount of nitrogenous waste products to be eliminated and the organs of elimination would be subjected to an undue strain. A varied diet is therefore best. Along with meats there should be eaten fruits, vegetables and cereals in order to supply the carbohydrates which afford the main source of our energy. Milk, the sole food of the infant, contains proteins, fats and carbohydrates in about the proper proportions. Most of the proteins of milk may by proper methods be converted into cheese. After the milk stands the fat rises as cream to the surface and may be made into butter. The carbohydrates are mainly in the form of milk sugar or lactose.

The kind and amount of food needed depends upon climate, habits of life and the peculiarities of the individual person. The Esquimaux may eat much more fats and carbohydrates than would be good for men in a temperate climate because they need food that can be utilized as fuel. A man at hard labor may likewise utilize more of these kinds of food than the gentleman of leisure. Perhaps most of us eat rather more than is necessary, and it is certain that overeating is a fruitful source not only of disturbances of digestion, but of various other bodily disorders. When we eat too much, consider what must happen. The digestive organs are overstimulated in the effort to dispose of the extra food. Constipation frequently follows. Materials which should have been got rid of undergo decomposition, producing injurious substances that are absorbed by the blood and poison the whole body. Excess of food commonly leads also to the torpidity of the liver. One of the functions of this much abused organ is the breaking up of various products resulting from the metabolism of the tissues. If the liver is sluggish, injurious substances may accumulate in the blood and produce very disagreeable feelings. Practically all of the carbohydrates that are absorbed from the stomach and intestines pass through the liver cells. Many people who eat too much rich candy or other sweets at all sorts of unseasonable times, and suffer from headache and general lassitude as a result, are apparently unaware that they bring these unpleasant consequences on themselves, by imposing upon the poor liver greater burdens than it can well endure.

The welfare of the liver is of especial value for the maintenance of health, because this organ performs so many indispensable functions. It destroys poisonous substances in the blood by converting them into less injurious materials. It eliminates various materials which are discharged through the bile duct into the intestine, while it secretes other substances whose presence in the intestine facilitates the absorption of food and checks the undue decomposition of waste matter. It also acts on dextrose (the substance into which carbohydrates are converted when they are absorbed into the blood), converting it into glycogen. There are other functions ascribed to the liver but these are the best known. The ills that arise from the maltreatment of the liver are numerous, and in many cases the patient never suspects that the liver is responsible for their production.

The fate of the different classes of food in the economy of the body may be summarized as follows:

Proteins.......
- produce energy;
- build up living tissue;
- may give rise to fat in the body.

Fats..........
- produce energy;
- are converted into bodily fat.

Carbohydrates..
- produce energy;
- are converted into fats;
- may be stored up as glycogen.

All foods produce energy for the performance of muscular work and the production of bodily heat. Carbohydrates and fats are chiefly energy producers and, although they may be stored for a time, they may be oxidized later as occasion demands, and hence used up for the production of energy in the end.

Besides the three classes of foods that have been described there are several other substances that are essential for the maintenance of life. Conspicuous among these is water, as it forms about $9/10$ of the material of the blood and about 59 per cent. of the substance of the body. Since water is constantly being given off from the body through the secretory activities of the kidneys, by the skin in the form of perspiration or sweat, and by the lungs in the breath, it must be supplied in considerable quantity. While more or less water is contained in all our foods, it is necessary to drink water or some beverage such as tea or coffee, etc., consisting mainly of water, in order to supply our bodily needs. A man may do without food for several days and in exceptional cases for some weeks, but he will not live nearly so long if he is deprived of water. Men under the hot and dry climate of the desert crave a large amount of water and soon succumb if they cannot obtain it.

In both our food and drink we consume small amounts of various kinds of salts. Comon table salt or sodium chloride, NaCl, is one of the most common of these, and while it occurs in small quantities in animal foods and often in drinking water, it is usually added as seasoning to much of the food we eat. Carbonate and sulphate of lime are common in drinking water, the so-called hard water containing an unusually large amount of one or both of these salts. These, with phosphate of lime, are used in the formation of bone, as well as in supplying calcium to other tissues of the body. Salts of potassium and magnesium are also important, and iron is required for the formation of the red coloring matter of the blood. While salts are needed only in small quantities, they are absolutely essential for the maintenance of life; the presence of several different kinds of salts is of even greater importance than the consumption of different classes of foods.

There are several substances which are consumed not so much on account of their value as food, but because they gratify the sense of taste. Spices add to the piquancy of various dishes, but they have practically no value as food and, although they may serve a good purpose by stimulating the appetite, some of them may produce bad effects, especially when taken in large amounts. There are several substances called *stimulants* which may or may not be of value as food. A stimulant is a substance that increases the metabolic activity of the organism. Frequently the excitement produced by the stimulant is followed by a period of depression in which the vital energy of the body is diminished. Contrasted with stimulants there are other substances called *narcotics* which have a quieting effect, often associated with pleasurable feeling. Common narcotics are tobacco, opium, chloroform and cocaine, whose influences will be discussed in a later section. Tea and coffee have a mild stimulating influence which they owe to alkaloids (thein and caffein) of similar if not identical composition. Both of these beverages, and especially tea, contain tannin which is extracted if water is allowed to stand too long on the grounds. Both tea and coffee are better when freshly made, not only because they contain less tannin, but because they retain more of their delicate aroma which soon disappears if either beverage is allowed to stand. Tannin is especially injurious to the stomach, and tea and coffee should be so prepared that most of this substance is not extracted. Tea and coffee may produce bad effects if used to excess, and there are some individuals in whom coffee, especially, produces disturbances of digestion, but perhaps the majority of people experience little harm from the use of these drinks. Chocolate, likewise a mild stimulant, is more nutritious than tea or coffee.

The most widely used of the beverages taken to gratify the sense of taste are the various drinks which contain alcohol; such as beer, wine, whisky, brandy and numerous others in almost endless variety. While it has been shown experimentally that alcohol is a food—*i.e.*, it may be oxidized in the body with the production of energy—it has practically little nutritive value as compared with other foods, and its other effects more than outweigh whatever nutritive value it may possess. The effect of alcohol on digestion when taken in large quantities is bad, as it gives rise to cirrhosis of the liver, inflammation of the stomach, dyspepsia, and a variety of other disturbances.

In smaller quantities its effect on digestion is less marked, and more difficult to determine. As people addicted to a moderate use of alcohol live, on the average, a less number of years than those who are temperate, the general effect of small quantities of alcohol on the body in general, and probably also on the digestive organs, is not good. People who take alcohol with their meals are apt to take more than is good for them, and it is quite certain that the digestion of normal, healthy persons is better without alcohol than with it. Those whose digestion is impaired should take alcohol, if at all, only as prescribed by a good physician.

CHAPTER XXVII

THE BLOOD AND CIRCULATION

When the soluble products of digestion diffuse through the walls of the alimentary canal they pass directly or indirectly into the blood, and are carried by this fluid to all parts of the body. The blood comprises about $1/13$ of the weight of the body, although it varies greatly in amount at different times and with different people. It is composed of a fluid called *plasma* and numerous very minute *corpuscles* which are so small that as many as 5,000,000 are estimated to occur in a cubic millimeter. The plasma of blood is a very complex fluid. It contains many food products, proteins, fats, sugar and various salts which may be taken up by the cells of the body; it contains also the waste matter derived from the destructive metabolism of cells. And there are various other substances in it which have a number of different functions. Blood is the great medium of transport of food, oxygen and waste. Each cell takes out of it the material needed for its life and gives off into it the broken down products of its vital activity. The blood has to keep in circulation in order to supply all the parts of the body which are dependent on it. And in the normal life of man it never stops for a moment from before birth to old age.

When blood is withdrawn from its vessels it has the curious property of forming a solid, jelly-like mass, the *clot*. This clot is composed of a substance called *fibrin*, a form of protein which is supposed to be derived from a soluble substance called *fibrinogen* by a process of fermentation. Clotting is therefore analogous to the formation of cheese (casein) from a protein which was previously dissolved in the milk. Exposure to air and especially contact with solid objects causes the blood to clot, and the process may be hastened by adding fine powder to the blood or beating it with a stick. The clotting of blood performs the very useful function of checking bleeding; otherwise bleeding would be very difficult to control and even slight wounds might produce fatal results. The yellowish fluid which remains after the fibrin has been removed by clotting is called *serum*.

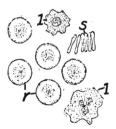

FIG. 193.—Human blood corpuscles. *l*, leucocytes or white corpuscles; *r*, red corpuscles; *s*, red corpuscles seen on edge.

The corpuscles of the blood which are true cells are of two kinds, the *red* and the *white corpuscles* or *leucocytes*. The red corpuscles are round biconcave disks; they do not contain a nucleus, although a nucleus occurs in the early stages of their development. Their most important ingredient is *hæmoglobin*, the substance which gives the blood its red color. Hæmoglobin is a protein containing iron and having the property of combining readily with oxygen, and also of giving up its oxygen again with equal readiness. This curious property enables hæmoglobin to perform its important function of a carrier of oxygen, since when oxygen is abundantly supplied to the blood as it circulates through the lungs the hæmoglobin becomes oxidized; whereas when the blood passes into a region where the cells of the body use all the oxygen that is available the hæmoglobin gives up its oxygen or, as the chemists would say, becomes reduced. Hæmoglobin combined with oxygen is red in color, while the reduced hæmoglobin is bluish;

a fact which accounts for the difference between the red color of blood fresh from the lungs, and the bluish color of blood in veins from other parts of the body. Blue blood is therefore blood with little oxygen.

While the red corpuscles are specialized for the function of carrying oxygen, the white cells or leucocytes perform very different functions. These leucocytes are very active cells with irregular, changing form. They have the property of creeping about much like Amœbas which indeed they closely resemble in form and general behavior. They engulf and digest many foreign materials and they are known to devour bacteria and other minute organisms that gain access to the blood. By means of this property they guard the body against many disease germs that might otherwise have an opportunity for unrestricted multiplication. This appetite of the leucocytes for bacteria renders the body more or less immune to various diseases. Leucocytes tend to congregate around centers of bacterial infection, and they may even pass through the walls of delicate blood vessels and creep about in the tissues, especially in regions of injury or bacterial invasions. If small tubes containing cultures of certain bacteria are introduced under the skin of a rabbit it is found that leucocytes creep into the tubes, while other tubes similarly prepared, but containing no bacteria are not entered. Apparently, therefore, the leucocytes are drawn into the tubes with bacteria on account of the fact that the bacteria produce some substances that attract these wandering cells. In regions where swelling occurs there are generally large numbers of leucocytes. Pus, which is a common product of inflammation, is composed largely of leucocytes together with broken down cells of other kinds. Leucocytes wander through the walls of the alimentary canal and they may also be found in the mouth. The whitish color of the coated or furred tongue is caused mainly by these outwandering cells which are especially abundant under certain abnormal conditions.

The organ that keeps the blood in constant circulation is the heart, which is situated near the middle of the chest with its pointed lower end lying on the left side between the fifth and sixth ribs, where we can plainly feel its beating. To understand how the beating of the heart causes the flow of blood we must study its inner mechanism. The heart is composed mainly of muscular fibers whose periodic contraction affords the energy for propelling the blood. It is divided into four chambers, the two *auricles* above and the two *ventricles* below. The auricle and ventricle of one side are separated from the corresponding chambers on the other side by a median partition which completely shuts off all communication between the two sides. Each auricle communicates with the ventricle of the same side by a valve which allows blood to pass from the auricle into the ventricle, but prevents its flow in the reverse direction. The ventricles are connected with the outgoing blood vessels or *arteries*, while the auricles receive the incoming vessels or *veins*. The muscular walls of the ventricles are considerably thicker than those of the auricles as they have to force the blood through the organs of the body, whereas the auricles simply propel the blood they receive into the ventricles. The valves which are present where the arte-

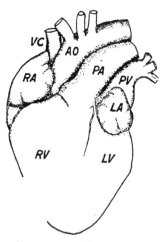

FIG. 194.—The heart seen from in front. *AO*, aorta; *LA*, left auricle; *LV*, left ventricle; *PA*, pulmonary artery; *PV*, pulmonary veins; *RA*, right auricle; *RV*, right ventricle; *VC*, superior vena cava.

ries leave the heart prevent the backward flow of the arterial blood.

There are two main arterial trunks, (1) the *pulmonary* which leads from the right ventricle and soon divides nto the pulmonary arteries which supply the lungs; and (2) the *aorta* which passes from the left ventricle and carries the blood which supplies the greater part of the body. Corresponding to the arteries which pass out from the ventricles are two sets of veins emptying into the auricles,

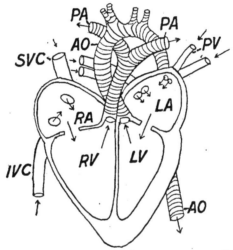

FIG. 195.—Diagram of inside of heart. *AO*, aorta; *IVC*, inferior vena cava; *LA*, left auricle; *LV*, left ventricle; *PA*, pulmonary arteries; *PV*, pulmonary veins; *RA*, right auricle; *RV*, right ventricle; *SVC*, superior vena cava.

(1) the *pulmonary veins* which bring blood from the lungs to the left auricle, and the two *venæ cavæ* which discharge blood from the general circulation into the right auricle. There are two systems of circulation connected with the heart, the *pulmonary* which carries blood to and from the lungs, and the *systemic* which carries blood to and from the rest of the body. Nevertheless the same blood must pass through both systems.

In order to illustrate its course let us start with the

blood as it passes from the left ventricle through the aorta. From this large vessel it may flow into any of the branch arteries which supply the arms, legs, liver or any of the numerous organs of the body. As it passes through the fine capillaries it is collected into veins which ultimately lead into the two venæ cavæ which discharge into the right auricle. From here it passes through a valve into the right ventricle whence it is forced out through another valve into the trunk that supplies the pulmonary arteries leading to the lungs. After passing through the capillaries of the lungs, it is returned by the pulmonary veins to the left auricle, whence it flows through a valve into the left ventricle, thus completing the entire circuit of both the pulmonary and the systemic circulations.

The vessels which carry the blood to and from the organs of the body fall into three classes, the *arteries,* carrying the blood from the heart; the *veins* which return the blood to the heart; and the minute *capillaries* which connect the arteries with the veins. Both arteries and veins have muscular walls, but the walls are thicker in the arteries in adaptation to the greater pressure to which they are subjected by the pumping action of the heart. As the arterial walls are elastic they expand somewhat as blood is forced into them by each contraction of the ventricles. This periodic expansion forms the pulse which can plainly be felt in the arteries of the wrist and neck. The rapidity of the pulse, normally about 75 beats per minute, is an index of the activity of the heart. As one may readily demonstrate upon himself, the pulse is quickened by exercise and excitement; it is also quickened during a fever, its rate often affording the physician a clue to the condition of the patient. In other conditions of illness the pulse may be unusually slow and weak.

As the arteries branch into smaller and smaller vessels

their walls become thinner and finally lose their muscular coat altogether as they pass into the minute capillaries. If the web of a frog's foot be examined with a compound microscope one may witness the fascinating spectacle of the veins, arteries and capillaries with their flowing currents of blood, and follow the red and white corpuscles as they pass, single file, in their course from the arteries to the veins. It is through the thin and delicate walls of the capillaries that the principal exchanges of material

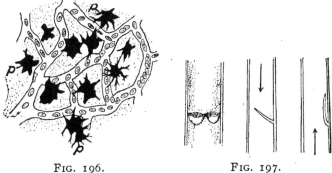

FIG. 196. FIG. 197.

FIG. 196.—Diagram of circulation through the capillaries of a frog's foot showing corpuscles in capillaries. *p*, pigment cells.

FIG. 197.—Valves in a vein. *A*, valves as they appear when a vein is cut open; *B*, section of vein to show closing of valve to prevent backward flow of the blood; *C*, section showing position of valve when blood is flowing normally.

occur between the blood and the tissues. Food and oxygen diffuse from the blood into the tissues and carbon dioxide and other waste products diffuse from the tissues into the blood. The white corpuscles may pass through these delicate walls also, as may sometimes be seen in the web of the frog's foot. The veins, whose function it is to carry the blood back to the heart, are furnished with cup-shaped valves which allow blood to flow past them toward the heart while they fill and block its passage if it should be forced in the opposite direction. Many veins lie nearer the surface than the arteries and they may be easily seen in such places as the back of the hand. If the wrist is grasped tightly so as to check the return flow of the blood the veins of the hand may be seen to swell.

Most of the blood sent to the stomach and intestines is not returned at once toward the heart, but is collected by the *portal vein* which distributes it to the liver. The portal circulation is therefore peculiar in that it both begins and ends in capillaries. Much of the food material absorbed from the stomach and intestines passes into the capillaries that lead to the portal vein. Sugar in the form of dextrose passes into the portal circulation and is converted into glycogen in the cells of the liver.

Besides the blood, the body has a similar but colorless fluid, the *lymph*. The lymph contains leucocytes, but no red corpuscles. It flows in numerous vessels, the *lymphatics*, which pour their fluid into the blood. Lymph vessels are abundantly supplied to the intestines where they are called *lacteals* on account of the milky appearance (lac, milk) of the lymph in this region caused by the presence of fatty substances absorbed from the intestine. The intestinal lymphatics converge into the *thoracic duct* which empties into a vein near the left side of the neck.

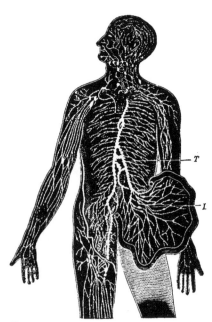

FIG. 198.—Lymphatic system. *L*, lacteals; *T*, thoracic duct.

The lymph does not have a complete circulation like the blood. It carries material from the surface of the body and the various organs into the blood vessels. It is replenished by fluid which diffuses from the finer blood vessels into the surrounding tissues. One of the chief functions of lymph vessels is the absorption and transfer of materials into the blood system.

Closely associated with the lymph vessels are the lymph glands. Some of these may be felt in the groins or under the jaw. They are centers in which leucocytes arise, and they often become enlarged in disease. Infections often travel from their point of origin, along the course of the lymph vessels and frequently cause swelling and suppuration of the neighboring lymph glands. The tonsils contain much lymphatic tissue, they are very apt to become infected and often have to be removed because the infection may spread from these organs and produce a variety of ill effects.

The organs of circulation are to a certain extent under the control of the nervous system. The peripheral arteries may contract upon nervous stimulation, as when one suddenly grows pale, or they may become relaxed and filled with blood under other circumstances, as in blushing. The exercise of a part causes it to receive an increased blood supply, and after a meal there is an unusually large amount of blood sent to the organs of digestion. Hard thinking brings an increased blood flow to the brain. Serious mental work after a heavy meal when the blood is occupied with the business of digestion is therefore an uphill task, as doubtless most of you have already found out.

The proper working of the organs of circulation is very important for the maintenance of physical vigor. The heart may be weakened by over exercise, but moderate exercise strengthens it as well as gives tone to the blood vessels. The heart is more frequently injured by bad habits and disease than by overwork. The continued use of alcohol in excess generally leads to cardiac weakness. Under the stimulus of a moderate amount of alcohol the peripheral blood vessels become dilated; more blood goes to the surface, and a feeling of warmth may be produced which, however, soon passes away. Alcohol probably produces disorders of circulation mainly through its influence on the nervous system, thereby causing an impairment of the proper nervous control of the heart and blood vessels. Tobacco, especially when used by the young, leads to heart weakness and palpitation, the "tobacco heart" being a frequent result of the use of tobacco in excess.

CHAPTER XXVIII

RESPIRATION

If we hold our breath for a short time we soon experience a sense of discomfort which increases the longer our breathing is interrupted, until it becomes quite intolerable. While we can go without water for some time and without food for a much longer time, we would very quickly succumb (it would be a matter of a very few minutes at best) if deprived of air. The element in the air upon which we are so closely dependent for our life is oxygen, the nitrogen being simply so much inert substance that plays no important part in respiration. While we may be prone to think of breathing or respiration as drawing air into the lungs and forcing it out again, these processes are merely subsidiary to the essential part of respiration which consists in the assimilation of oxygen and the giving off of carbon dioxide. Liquids tend to absorb gases when the latter are present in considerable quantities, and they give off gases when there is nothing to check their escape. If a liquid such as water is separated from the air by a permeable membrane it may absorb air through it and give off any gas which it may contain in excess. Blood has this property, like other liquids, and it has an especial aptitude for absorbing and giving off unusually large amounts of oxygen and carbon dioxide, the chief gases concerned in respiration. Gases have the general property of tending to become uniformly distributed. If a bladder is filled with oxygen and suspended in an atmosphere of carbon dioxide the oxygen will diffuse out of the bladder while the carbon dioxide will diffuse into it until the two gases are evenly distributed on either side of the membranous wall.

The ability of the blood to take up large quantities of oxygen is dependent upon the fact that the hæmoglobin of the red corpuscles forms a chemical combination with this element. When supplied with oxygen the blood turns red as may be demonstrated by shaking up bluish venous blood with oxygen or even with air, when it takes on a bright reddish color. Similarly red blood becomes bluish when shaken up in an atmosphere of carbon dioxide. As we have already seen, the blood returning from the lungs where it becomes exposed to the air is red in color and remains red in the arteries that supply the organs of the body. After it has passed through the capillaries and is returned through the veins it acquires a bluish tint. As chemical tests show, it has lost a considerable part of its oxygen and has received a larger amount of carbon dioxide. The constant need for oxygen to keep up the life of the tissue and the necessity for the removal of carbon dioxide, which is produced where vital changes are going on, make it so very important that the activity of breathing should go on unchecked.

While we speak of the lungs as the organs of respiration it must be remembered that respiration occurs in all the cells of the body. They all take oxygen from the blood and give off carbon dioxide. This process is often distinguished as *internal* or *tissue respiration*. But as the blood circulates through the lungs it absorbs oxygen from the air and gives off its carbon dioxide, a process which we may designate as *external respiration*. The blood acts as a carrier between the tissues and the respiratory organs, a function, as we have seen, for which it is admirably adapted. Oxidation is essentially a process of burning, and the consumption of oxygen by the tissues affords an important source of our bodily heat. And the main products of this oxidation, water and carbon dioxide, are precisely those which are formed by the burning of a candle or a stick of wood.

As the lungs are the principal organs from which the blood gets its oxygen we may now consider their structure,

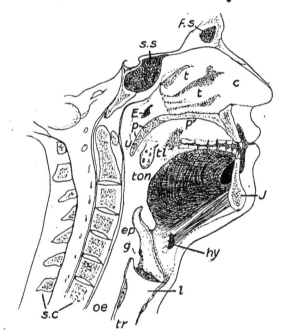

Fig. 199.—Median section through nasal cavity, mouth and throat. *C*, cavity of nose; *E*, opening of Eustachian tube from the ear; *ep*, epiglottis; *f.s*, frontal sinus; *g*, glottis; *hy*, hyoid bone; *j*, lower jaw; *l*, larynx; *oe*, esophagus; *p′*, hard palate; *p*, soft palate ending posteriorly in the uvula, *u*; *s.c*, bones of spinal column; *s.s*, sphenoidal sinus; *t, t*, turbinated bones; *tl*, tonsil; *ton*, tongue; *tr.* trachea; *u*, uvula.

and see how they are adapted to the performance of their functions. In order to enter the lungs the air has to flow through a number of passages. First it is taken into the nasal cavity where it is exposed to a wide surface of moist mucous membrane. Here the air is not only warmed before passing to the lungs, but dust and other particles are caught and prevented from interfering with respiration. Posteriorly the nasal cavity leads to a space called the *pharynx;* at the lower end of this is a cartilaginous box,

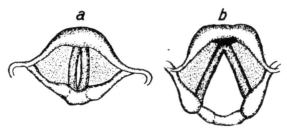

FIG. 200.—Vocal cords. *a*, closed; *b*, open.

the *larynx,* which you can feel in the front part of your throat (Adam's apple). The opening of the larynx lies just in front of the opening of the esophagus, but it may be covered by a fleshy lid, the *epiglottis,* which normally closes it during the act of swallowing. Sometimes food "goes down the wrong way" when it sets up the act of coughing by which it is usually expelled.

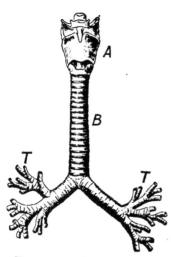

FIG. 201.—Larynx *A* and windpipe *B* with its branches; *T,* bronchial tubes.

The larynx contains the vocal cords whose function it is to produce the voice, and it leads below to a tube called the *trachea* or *windpipe* through which air passes to the lungs. The trachea is furnished with a series of C-shaped cartilages or nearly complete rings which keep its walls from collapsing and thus impeding the passage of air. The inner lining consists of mucous membrane containing many cells with fine cilia whose constant beating creates a current toward the mouth. Fine particles which may lodge in the trachea are therefore carried outside instead of being allowed to accumulate in the lungs. Nature has furnished us with a number of admirable contrivances by which the lungs are shielded from injury. The broad nasal cavity for warming the air and collecting dust, the epiglottis which closes the opening to the air passages at the very moment when materials are apt to pass into them, the tracheal cartilages to keep open the trachea and thus to insure access of air to the lungs, and the fine tracheal cilia beating in the right direction to carry away offending particles—all these structures act so as to provide the lungs with air devoid of solid matter.

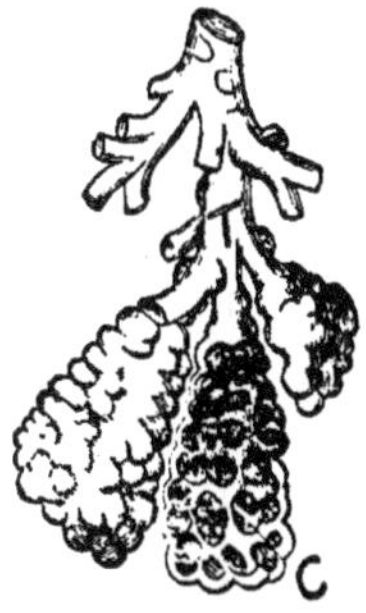

At its lower end the trachea divides into two *bronchi*, one for each lung, and these two tubes subdivide into smaller and smaller ones. The final subdivisions lead to minute pockets, the *air cells*, the walls of which are exceedingly thin and abundantly furnished with capillary blood vessels. A large surface is thus provided in which the blood is brought into intimate contact with air, the thin walls by which the two are separated facilitating the exchange of oxygen and carbon dioxide which, as we have seen, is the essential function of organs of respiration. The total surface of the numerous air cells is estimated to be about 15,000 square feet, an area equal to the floor space of a fair-sized dwelling.

FIG. 202.—Lobules of lung, the lower one cut open showing the air cells, C; bronchial tubes, T.

The lungs are fairly large organs, pinkish in color and of very spongy texture, and they fill most of the chest which is not occupied by the heart. They are surrounded by a double membrane, the *pleura*, one layer of which is closely applied to the lungs while the other forms the inner lining of the chest. The space between the two layers contains a fluid which serves to prevent friction from the constant movement of the chest in breathing. Pleurisy is a disease due to the inflammation of the pleura.

We have now to consider how the air goes in and out of the lungs. The ribs enclosing the chest are capable of more or less movement by means of various muscles which are attached to them. By raising up the sternum or bone to which the upper ribs are attached in front, and by spreading the lower free ribs laterally the chest becomes enlarged, and, as the lungs expand at the same time, air tends to rush in from without to fill the extra space—a process which goes by the name of *inspiration* (literally breathing in). At the lower side of the cavity of the chest is a broad muscular sheet, the *diaphragm*, forming a complete septum across the body. Usually the diaphragm is arched upward in the center, but when its muscular fibers contract they cause this arch to be flattened downward, thus further enlarging the cavity of the chest.

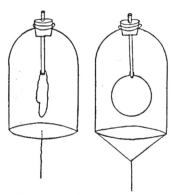

FIG. 203.—Diagram to illustrate the effect of depressing the diaphragm on the air content of the lungs.

Relaxation of the diaphragm and contraction of the muscles that draw in the ribs cause the air to be forced out of the lungs—a process which goes by the name of *expiration* (literally breathing out). The alternate movements of inspiration and expiration, while to a certain extent under the control of our will, go on for the most part quite unconsciously. Their rapidity varies with age, being greater in young children, gradually becoming slower as they grow older. It also varies with exercise like the rapidity of the pulse and for much the same reason. When we work hard more oxygen is consumed and more carbon dioxide is produced and the lungs, thus compelled to become more active, require air to be pumped in and out with greater rapidity. At the same time the blood must be hurried through the lung capillaries in order that it get rid of its surplus of carbon dioxide and receive the requisite supply of oxygen for the increased demands of the tissues. This is why when we run hard for a train we find ourselves panting for breath and our pulse beating wildly; we become heated too as a result of the increased metabolism that goes on in our muscles.

As respiration is so intimately associated with the maintenance of life it is especially important that we be supplied with an abundance of fresh air. If we live in small rooms into which air from the outside cannot freely enter, the air supply soon becomes contaminated with carbon dioxide and other noxious products. We soon experience a sense of lassitude and depression, and if we habitually live under such conditions our general health will inevitably become impaired. Colds, consumption and various other diseases are more readily contracted by persons who live in impure air. Rooms should be ventilated so as to secure a free circulation of air without exposing their inmates to cold draughts. In order to breathe properly the chest should not be allowed to become deformed as it is in many people with stooping shoulders. Students who sit at desks which are too low are particularly liable to this deformity. The chest capacity is diminished and the whole body consequently suffers.

Many women through the absurd habit of tight lacing compress their bodies so that the lower part of the chest, which normally expands most in breathing, can scarcely expand at all. Consequently they breathe mostly with the upper part of the chest and greatly diminish the supply of air which the lungs require. Not only this, but the abdominal organs are displaced thus giving rise to various other harmful effects.

Respiration is a function common to all organisms without exception. In many of the primitive animals respiration takes place through the whole surface of the body. Only in higher forms are there specialized organs of respiration and these commonly consist of structures by means of which a relatively large surface can be brought into contact with oxygen. In aquatic forms this surface is usually in the form of outgrowths, such as filaments or plates, the walls of which are very thin so as to permit the free interchange of gases. Gills of various forms are to be met with in many worms, mollusks, crustaceans, several aquatic insects, in all fishes and usually in the young of amphibians. The oxygen which the gills utilize is, as a rule, the free oxygen which is dissolved in the water. If animals are placed in water from which the oxygen has been driven off they will die of suffocation.

In land animals the increase of surface for exchange of gases is usua ly obtained not by outgrowths, such as gills, but by ingrowths, such as the tracheal tubes of insects or the lungs of higher vertebrates. As the thin membrane that separates the blood from the oxygen must be kept moist for the proper transfer of gases the exposure of delicate gills to dry air, to say noth'ng of dust and dirt, would be very disadvantageous. In all higher land animals Nature has safely located the organs of respiration within the body where their delicate surfaces are always moist and furnished with an abundant supply of blood. All of the varied organs of respiration in the animal kingdom are devices for securing essentially the same end, whether they are gills, tracheal tubes, lungs, or simply the general surface of the body as in the Hydra and earthworm.

CHAPTER XXIX

EXCRETION

As living matter is constantly being torn down and built up, the removal of waste is as important as supplying food or oxygen. The process of getting rid of waste materials is known as *excretion*. It is performed by several organs such as the kidneys, the liver and the skin, each of which carries on its own peculiar kind of excretory activity. Every cell gives off waste into the blood just as every cell respires and assimilates food. A part of this waste is CO_2 and is gotten rid of mainly through the lungs, while other waste materials are solid and escape from the body only by the medium of water, in which they become dissolved. As the lungs are specialized to get rid of the gaseous waste, so other organs are peculiarly adapted to get rid of other forms of waste which are given off into the blood by the cells of the body. Chief among these organs of excretion are the kidneys, two reddish organs one on either side of the spinal column just below and behind the stomach. Each kidney receives a large artery from the aorta (the *renal artery*) and gives off a large vein (the *renal vein*) that joins the inferior vena cava. Passing from each kidney is a duct called the *ureter;* the two ureters pass downward to connect with the *bladder* which is a thin-walled, elastic sac which serves as a reservoir for the storage of the fluid secreted by the kidneys.

The kidney is a gland consisting mainly of numerous *uriniferous tubules.* Each of these structures begins in a

Malpighian corpuscle which consists of a thin-walled capsule surrounding a knot of blood vessels, the *glomerulus*. These tubules after a tortuous course lead to a cavity which is connected with the ureter. The uriniferous tubules which we may call the drain pipes of the body remove a considerable quantity of water and various salts from the blood, but their most important function is the elimination of urea which is a nitrogenous compound re-

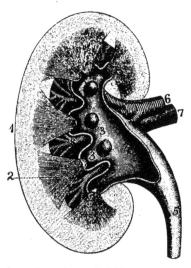

FIG. 204.—Vertical section through kidney. 2, pyramid showing the renal tubules opening at the papillæ, 3; 4, pelvis or cavity of kidney; 5, ureter; 6, renal artery; 7, renal vein; 8, branches of these blood vessels. (After Leidy.)

sulting from the breaking down of proteins. The elimination of water affords the means of getting rid of solid waste products in the blood, because these substances must be in a state of solution in order to be discharged from the body. Most people would be better off if they drank more water so as to wash their protoplasm free from its various injurious impurities. We are all poisonous to ourselves. When disease of the kidneys impairs their activities poisons accumulate in the body and death may result. Some of the diseases of the kidneys such as chronic Bright's disease are among the most fatal of human maladies. One of the most potent causes of kidney diseases is a'cohol, because so much blood passes through these organs that the secretory cells are especially exposed to the influence of this drug. Excessive beer drinkers are especially prone to kidney disease not only on account of the alcohol they consume, but because of the strain imposed on the kidneys to remove the large amount of fluid added to the blood.

The liver is also an important organ of excretion not only on account of the substances it discharges into the intestine through the bile duct, but because it acts on various products of protein metabolism, converting them into urea, in which form they are given off into the blood to be removed by the kidneys. Certain substances, especially salts, are thrown off in the perspiration of the skin.

CHAPTER XXX

INTERNAL SECRETIONS AND THE DUCTLESS GLANDS

We have already given several examples of ordinary secretion in which a gland pours out certain substances to the outside through a duct. While salts, urea and other materials are taken as such out of the blood and passed through the glands unchanged, in many other cases the substances that are discharged are manufactured by the glands themselves. Such substances are found in the saliva, gastric juice and bile. Hydrochloric acid and pepsin do not occur as such in the blood, but are made in the cells of the gastric glands. Secretion in these cases, therefore, does not consist merely in filtering out materials that are present in the body, but in the formation and discharge of new compounds. There are many cases in which the compounds formed by an organ are not discharged to the outside but are given off into the blood. This process is known as *internal secretion*. We have already met with one example of this in the urea which is formed in the liver out of various products of protein metabolism and given off into the blood to be eliminated from the body by the kidneys. This substance, therefore, is an internal secretion of the liver and an external secretion of a quite different organ.

Many organs which produce internal secretions have no outlet and hence are known as *ductless glands*. The function of most of the ductless glands was for a long time unknown, but it is now well established that some of these organs produce substances that are essential to the maintenance of life. One of these is the *thyroid gland* which is situated in the fore part of the neck under the larynx. Complete removal of the thyroid results in death, which, however, may be obviated if the substance secreted by this organ is given to the patient. This substance contains iodine and is now a remedy that is regularly kept at drug stores. There is a peculiar disease called cretinism, caused by impairment of the thyroid, which is associated with certain bodily abnormalities and especially lack of mental development. Administration of thyroid extract has a wonderful curative power in such cases, and it often converts a child who is backward to the point of idiocy into a bright and happy little person. There was simply supplied the internal secretion necessary for normal development which its own body failed to furnish in sufficient amount. Another disease, goiter, is caused by an abnormal enlargement of the thyroid.

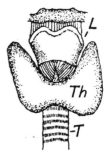

FIG. 205.—*Th*, thyroid gland; *L*, larynx; *T*, trachea.

Other glands that produce a substance essential to life are the *adrenals* or *suprarenal bodies*, small glands immediately over the kidneys. Preparations made from these glands (adrenalin, epinephrin) have the property of contracting the blood vessels of a part to which they are applied and they are therefore used in minor surgical operations to prevent bleeding.

The pancreas in addition to the pancreatic juice discharged into the intestine secretes a substance that has to do directly or indirectly with the metabolism of sugar. Removal of the pancreas results in the accumulation of sugar in the blood (diabetes) and eventually in death. If, however, a small part of the pancreas is grafted in some other part of the body death may be prevented. Although there is no longer any secretion of pancreatic juice, the pancreas continues its internal secretion which is the most important of its functions. Other organs have important internal secretions but they cannot be described here.

CHAPTER XXXI

THE SKIN

The most obvious function of our skin is that of protecting the organs which it covers, but it serves also as an organ of excretion, a medium for the regulation of bodily temperature, and a sensory surface adapted to receive many kinds of impressions from the outer world. A few facts about the structure of the skin will help us to understand how it performs its functions.

The skin is composed of two layers, a deeper one the *corium*, and an outer *cuticle* or *epidermis*, which consists of epithelial cells without blood vessels or nerves. The outer cells of the cuticle are dead and as they are shed or rubbed off, they are continually replaced by cells from beneath. When a blister is formed the cuticle is lifted away from the corium by serum which exudes between these two layers. The corium is a relatively thick layer of connective tissue containing muscle fibers, blood vessels, glands, the end organs of nerves, and various other structures. The deeper part of the cuticle contains brownish pigment which in colored races is especially abundant and which in ourselves is increased when we become freckled or tanned.

Among the most characteristic structures of the skin are the sweat glands which are coiled tubes whose ducts open through fine pores at the surface of the cuticle. With a good hand lens these openings may be seen in the palm of the hand. As many as 2,000,000 of them have been estimated to occur over the entire body. Sebaceous or oil glands are present wherever hair occurs which is over most of the body except the palms of the hands and the soles of the feet. Over most of the surface of the body these hairs are inconspicuous; they represent the mere remnant of the coat that once covered our hairy ancestors. There are usually two oil glands to each hair follicle and their secretion serves to keep the hair oily. The hair of the head may be supplied with more oil if the glands are stimulated by massaging the scalp or vigorously brushing the hair.

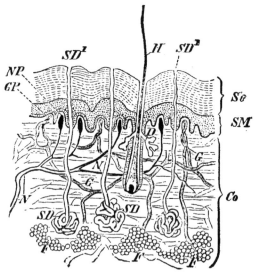

Fig. 206.—Section of skin of man. (From Wiedersheim.) *Co*, derma (corium); *D*, oil gland; *F*, fat; *G*, blood-vessels; *GP*, vascular papilla; *H*, hair; *N*, nerves; *NP*, nerve papilla; *Sc*, stratum corneum; *SD*, *SD*¹, sweat gland and duct; *SM*, stratum Malpighii.

Each hair grows from a little papilla at the bottom of the hair follicle, where there are many small blood vessels and nerves. Sometimes hairs fall out, but if the papilla is uninjured they may be grown again. Hairs contain pigment which gives them different colors in different individuals, but all kinds of hairs tend to become gray with old age owing to the central part of the hairs being filled with air. When the scalp is ill nourished the hairs that fall out may not be replaced and the head becomes bald. Then the afflicted person generally resorts to hair tonics most of which are utterly useless. The best way to avoid the dreadful fate of having a shiny bald head is by washing the hair occasionally with soap and water, by brushing the hair well and keeping up a good circulation in the scalp.

We have said that one function of the skin is to regulate the temperature of the body. How does it do this? Most of the lower animals are, as we say, cold blooded. Their temperature goes up and down with the changing temperature of their surroundings. But we are endowed with a remarkable system of heat regulation which is so perfected that between the heat of summer and the extreme cold of winter our bodily temperature scarcely fluctuates more than a degree. Of course what keeps up our temperature is the burning of fats, carbohydrates and proteins in our tissues, but our skin regulates the rapidity with which heat is allowed to escape, and in this way keeps our temperature uniform. Our blood vessels are under the control of nerves which regulate their diameter and thus control the amount of blood that passes through them. When the nerves of the skin are stimulated by cold they generate impulses in the nerves supplying the blood vessels of the skin causing these blood vessels to contract. The blood is driven from the skin and hence does not radiate heat so rapidly to the outside. When the blood is too warm either from surrounding heat or from exercise there is an enlargement of the blood vessels of the skin. More heat is radiated and at the same time the sweat glands secrete more perspiration which is poured out at the surface where it evaporates. Evaporation always produces a lowering of temperature. Wet your fingers and then wave your hand through the air so as to make the water evaporate more rapidly and you will feel a sensation of coolness. The more perspiration is evaporated at the surface the greater the cooling effect; thus by means of the changes in the circulation of the skin and the activity of its glands our skin is able to act as a self-regulating mechanism, keeping the temperature always at a certain point.

In order that our skin can perform its functions properly, especially the regulation of temperature, it must be kept clean so that its pores are prevented from becoming clogged. We not only acquire dirt from the outside, but we sweat it out from within, and the waste matters in the form of salts, and various impurities accumulate, some of them to undergo decomposition to the distress of our associates if not of ourselves.

But bathing is desirable not only to get rid of dirt, but as a tonic to the skin and incidentally to the whole body. The cold dash or shower for a moderately robust person tones up the blood vessels of the skin and renders him much less liable to colds and various other forms of infection. If after taking a cold bath a person does not respond, after a vigorous rubbing, so as to feel an exhilarating glow, he had better confine himself to warmer water.

CHAPTER XXXII

THE SKELETON AND THE MUSCLES

The skeleton has for its general function the support and protection of the organs of the body. If we were to be deprived of our framework of bones we should collapse at once into an inert and flabby mass of flesh. What gives the bones the rigidity necessary for their functions is the presence of mineral constituents consisting chiefly of phosphate and carbonate of lime. These substances may be dissolved out of bone by means of nitric acid leaving a tough, gristly counterpart which may be readily bent or even tied in a knot. On the other hand, when bones are burned the tough animal matter is consumed, leaving only the mineral salts with which it was impregnated.

A part of our skeleton is formed of tough animal matter called *cartilage* which has but a very small quantity of mineral salts. Cartilage occurs when bones are growing and where flexibility is required, as at the ends of the ribs. At the joints the bones are bound together by very tough and inelastic bands of connective tissue called *ligaments*. In many cases the bones are joined so as to permit freedom of movement in one or more directions. We have the so-called hinge joints at the knee, and elbow, permitting a back and forth motion, and ball-and-socket joints at the hip and shoulder, enabling the limb to move in any direction. Movable joints are enclosed in a *synovial membrane* whose secretion, the synovial fluid, serves to lubricate the surfaces of the united bones. Many of the long bones like the longer bones of the arms and legs are hollow. With a given amount of material the greatest strength is secured

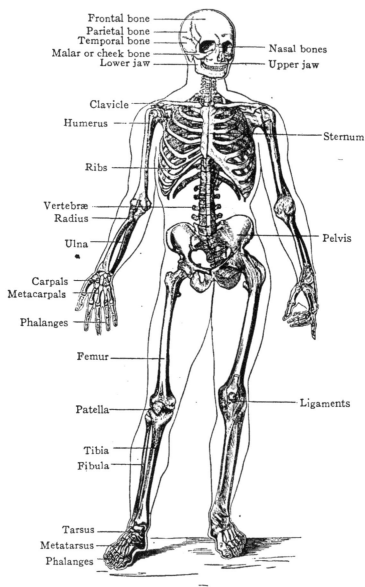

Fig. 207.—Skeleton.

if it is in the form of a hollow tube, and Nature has taken advantage of this principle in the formation of the skeleton.

But she has not wasted the space in the interior of the bones; they are filled with *marrow*, a substance which does not seem to have any particular use, but is really an essential organ of the body, for it forms the greater part of the red and white corpuscles of the blood.

We commonly have a few more than 200 bones in the body, the number being somewhat variable because some bones that are separate in youth become fused together in later life. In the skull many of the bones of the cranium, or part enclosing the brain, are united by sutures which dovetail together in such a way as to prevent them from becoming separated while at the same time affording a place where the bones can grow and thus give more space for the growing brain. As the bones of the head are very rigid, if they were fused together at the sutures the head could no longer enlarge. Growth takes place at the edges of the sutures.

The central axis or backbone of the skeleton (*spinal* or *vertebral column*) is composed of 24 bones, the *vertebræ*, separated from one another by elastic pads of cartilage. Inside of a canal running through the vertebræ is the spinal cord which connects with the brain above and sends off nerves between the vertebræ to nearly all parts of the body. Projections or processes of the vertebræ give attachments to ligaments and the muscles of the back.

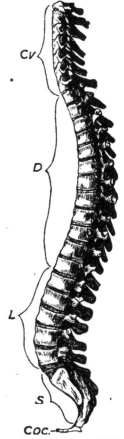

FIG. 208.—Side view of spinal column. *Cv*, cervical; *D*, dorsal; *L*, lumbar; *S*, sacral; *Coc.*, coccygeal vertebræ.

Attached by movable joints to the vertebræ of the trunk are the ribs. All but the two lower pairs are joined by cartilage to the flat *sternum* or breast bone which you can easily feel in the front of the chest. The last or floating ribs are free in front, thus permitting a greater expansion of the chest cavity.

Forming a support for the arms are the *scapula* or shoulder blade, and a narrow bone, the *clavicle*, which extends from the upper end of the sternum to the scapula near the articulation of the arm. Both these bones give attachment to muscles that move the arm and to certain other muscles of the neck and trunk. The upper bone of the arm, or *humerus*, is joined to the scapula by a ball-and-socket joint. At the elbow, one of the bones of the fore-arm, the *ulna*, is joined to the humerus by a hinge joint, while the other bone, the *radius*, which lies on the same side as the thumb is joined to the ulna in such a way as to permit it to rotate about the latter with the greatest freedom. The wrist composed of eight small bones, or *carpals*, is followed by the five *metacarpals* in the palm of the hand, and these give attachment to the *phalanges* of the thumb and fingers.

The hip bones which give attachment to the lower extremities are united into a solid arch, the *pelvis*, which is firmly joined to the fused vertebræ forming the *sacrum*. The large size of the pelvis is necessary for the attachment of the large muscles that move the legs as well as the various muscles of the trunk. There is a close similarity between the bones of the legs and those of the arms. Corresponding to the humerus is the *femur*, the head of which joins by a ball-and-socket joint to the pelvis. At the knee the *tibia* and *fibula*, corresponding respectively to the radius and the ulna, are united by a hinge joint to the femur. The carpals of the wrist are the representatives of the *tarsal* bones of the ankles, and these are followed by the *metatarsals*, and these again by the phalanges which are the skeletal elements of the toes. At the knee there is a small, rounded bone, the *patella* or *knee pan*.

The kind of material that is usually eaten as meat consists mainly of muscle, a tissue whose chief function is the production of movement. We are able to move only because our muscles have the property of shortening or contracting under the influence of stimuli. Grasp the upper arm with one hand while you bend the fore arm, and you can feel the large muscle in front of the humerus (the biceps) shorten and thicken. When you straighten the arm you can feel an opposed muscle (triceps) contract on the opposite side. In moving the arm the bones act as levers which are pulled this way and that by the muscles attached to them. Muscles never act by pushing, the opposed movement being always effected by the contraction of an antagonistic muscle. Accordingly we commonly find muscles in pairs the members of which have opposed functions. If a part is pulled in any one direction there must be some other muscle to pull it back again. Muscles are usually attached to bones, sometimes directly and sometimes by means of strong inelastic cords, or *tendons*. You can easily feel the tendons of your biceps muscle or the tendons of some of the muscles behind the knee or at the wrist.

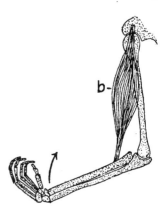

FIG. 209.—Diagram showing the action of the biceps muscle *b*, which when it contracts moves the fore arm in the direction of the arrow.

The contractile tissue is made up of fibers, which are

greatly elongated cells. In the so-called voluntary muscle, or muscle which we can control by the will, the fibers are marked with fine cross striations. In involuntary muscle the fibers are smaller and usually devoid of cross striations. Such muscular tissue occurs in the coatings of the stomach and intestines, in the walls of blood vessels and ducts, and in various other parts of the body. Involuntary muscle acts usually without our being aware of its contraction, and we have very little direct control over its activity. In fact it is best that we are not able to control it, because the wisest of us would not know how to direct its functioning.

FIG. 210.—Muscles and tendons of fore arm and hand.

Muscles increase in size and strength through exercise, and there is nothing so good as exercise in the open air, not only for muscles, but for the body in general. Muscular exercise involves stimulating to greater activity practically all parts of our organism. It brings increased appetite, better tone of the blood vessels, greater lung power, increased elimination of waste, better feelings and clearer thinking. When exercise is carried too far, however, we have the feeling of fatigue which is caused by substances in the blood that result from an excessive breaking down of living tissue, but when we have rested and the fatigue-producing substances are disposed of, Nature not only restores what has been lost, but she commonly adds a little more. Hence the strong biceps of the blacksmith, which is in striking contrast to the weak and flabby muscles of a sedentary clerk who never takes any vigorous exercise. And it is not for the sake of our muscles alone that we should exercise, but for the sake of our brains as well. For clear thinking depends upon a good body as much as does good digestion, and it is just as necessary for the student to attend to exercise in order to keep his body and brain in good working condition, as it is for the professional athlete.

CHAPTER XXIII

THE NERVOUS SYSTEM

We have seen that the movement of a part of our body is due to the contraction of muscle. But muscles contract because of a stimulus which they receive through a nerve. Without nervous impulses to initiate and direct movements the muscles would be inactive and useless. The nervous system is one which controls and regulates to a large degree the activities of the organs of the body, and to this end its nerve fibers extend into practically all parts of the organism. This system affords also the medium through which we feel various sensations. Almost all parts of our body are sensitive in one way or another, and, when we feel, it is due to the fact that some of our nerve fibers carry an impulse to the brain. The nervous system, and especially the brain, is very closely associated with the mind. It is only through this system that the mind is affected by influences acting on the body, or is able to produce bodily movement. When I burn my finger the impulse set up in the nerves supplying this organ passes to the spinal cord and thence to the brain. In consequence of this impulse I feel a very unpleasant sensation. When I pull away my hand, as I am pretty sure to do under the circumstances, there is an impulse sent out in the reverse direction, which is carried by a nerve to the muscles of the arm causing them to contract. Our nerve fibers perform functions very similar to those of telegraph and telephone wires which carry messages to and from a central station. The mind like the central operator

t work at cross purposes, so that we would be unable
omplish anything that we wanted to do.
order to see how the nervous system acts we need
n some knowledge of how it is formed. Our nerves
omposed of bundles of nerve fibers. Each fiber
ts of (1) a central core of nervous substance, the
ylinder; (2) an external very thin sheath; and (3)
st nerves a layer of fatty substance, the *medullary*
, between the axis and cylinder and the outer cover-
Each nerve fiber comes from a *nerve,* or *ganglion cell,*
terminates in various ways according to its function;
nerves supply muscles (*motor nerves*), some end in
, and others (*sensory nerves*) end in some sense organ
rve to carry sensory impulses to the central nervous
. Nerve cells commonly have several branches
f which may subdivide repeatedly. In the central
s system these branches meet and thus impulses
d by one cell may be conveyed to other cells and
be transmitted to remote parts of the body. Nerve
e frequently grouped together into masses called
which are found in various parts of our organism.
ge masses of nervous tissue making up the brain
inal cord have much the same composition as
, being composed of nerve cells and their branches,
r with blood and lymph vessels and a framework
ective tissue binding the whole together.
ervous system is composed of two principal parts,
bro-spinal system, and the *sympathetic.* The first
. the brain, the spinal cord, and the nerves that
from these central organs. Both brain and spinal
itain (1) *gray matter* which is composed largely
cells and their branching processes, and (2) *white*
hich is formed of nerve fibers with few or no cells.
pinal cord the gray matter is central, forming a

receives messages that come to the bra
sends out impulses which cause certain
contract; by this means we are enabl
way we desire in regard to objects in tl
Through the function of the nervous sys
impulses from one part of the body to an

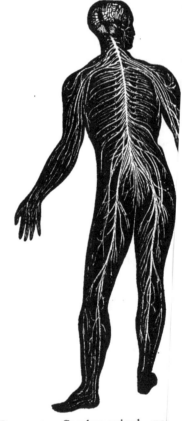

FIG. 211.—Cerebro-spinal ne

of different organs are made to har
gists say, become coördinated.
conducting system one arm might
the other arm, and each leg mig]
with the other one in walking,

migl
to ac
In
to ga
are
consi
axis
in m
sheat
ing.
and it
some
gland
and s
system
some
nervo
receiv
finally
cells a
ganglic
The la
and s
ganglia
togeth
of conr
The
the *cer*
include
proceed
cord co
of nerv
matter
In the

mass which in cross section has a certain resemblance to the letter H. The white matter that surrounds the gray mass is composed almost entirely of fibers running lengthwise of the cord. The spinal cord is lodged in a canal within the bones of the vertebral column and sends off a pair of nerves between each of the vertebræ. Each spinal nerve arises from the cord by two roots, a dorsal and a ventral, which penetrate the white matter and extend into the gray. The two roots soon unite and pass out of the spinal column as a single nerve. Each dorsal root is furnished with a ganglion, the cells of which give origin to the nerve fibers of the dorsal root and to others which proceed outward and form a part of the spinal nerve.

Experiment has shown that the fibers of the two roots have different functions. If the dorsal root is cut and the end in contact with the cord is stimulated a sensation of pain is felt that is referred to the part to which these particular nerve fibers are distributed. If the other cut end is stimulated no effect is apparent, and the part which the nerve supplies may be cut or burned without producing the least sensation. It is evident, therefore, that the nerves producing sensation pass into the cord through the dorsal root. Cutting a ventral root destroys all power of voluntary movement in the part supplied by the nerve in question, but the sensibility in that part remains undiminished. Stimulating the cut end of the root in contact with the cord produces no marked result while stimulating the outer cut end produces vigorous movement in the part to which the nerve is distributed. The ventral root contains the motor fibers, or those having to do with the production of motion.

The spinal cord is the center of many reflex acts. If the spinal cord of a frog is cut across near the middle and one of the hind toes is pinched, the leg is quickly withdrawn.

The brain of the animal has nothing to do with this act, as it no longer has any connection with the parts involved in the reaction. The impulse set up in certain sense cells of the foot travels toward the cord, passing through the dorsal root into the gray matter. Thence it passes out of the cord along the ventral or motor root to the muscles of the leg causing them to contract, thus drawing the leg

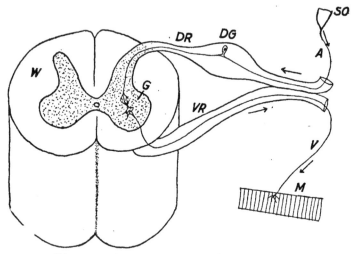

Fig. 212.—Diagram of a cross section of a spinal cord showing paths taken by nerve impulses in a simple reflex act. *A*, afferent or sensory nerve fiber entering the dorsal side of the cord; *DG*, dorsal ganglion; *DR*, dorsal root; *G*, gray matter of cord; *M*, muscle; *SO*, sense organ; *V*, branch of ventral or motor root of spinal nerve; *VR*, ventral root; *W*, white matter of cord.

away. If a drop of acid is put on the side of the same frog the hind foot on that side is brought forward to wipe away the irritating substance. These are but a few of the reflex acts which may be performed by the spinal cord. Such acts, although involuntary, are adaptive in that they are directed so as to perform some useful function. They are carried out in ourselves much as in the frog. If the sole of the foot is tickled the foot is quickly withdrawn even before we are aware of our action. We can check or prevent many of our spinal reflexes by means of impulses sent down from the brain, but none the less many such reflexes are often performed without our knowing it and they continue to be performed if the upper part of the cord is injured or paralyzed so that the lower part of the body is no longer under voluntary control. Reflex acts are performed by way of the brain and other nerve centers as well as by the

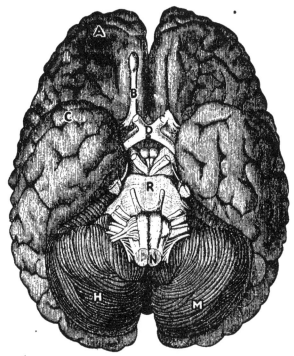

Fig. 213.—Lower side of brain. *A, C*, lobes of the cerebrum; *B*, olfactory tracts; *D*, crossing of the optic nerves; *H, M*, cerebellum; *N*, medulla; *R*, pons. (After Leidy.)

spinal cord. Coughing, sneezing and winking are some of these and they are often performed quite independently of our will.

Man is remarkable among animals for the great size of his brain. The largest part of the brain consists of the two *cerebral hemispheres* which are the organs especially concerned with voluntary action and the power of thought.

Though separated from one another by a deep median cleft or fissure they are broadly united by a mass of transverse fibers, the *corpus callosum*. The surface of the hemispheres is thrown into numerous folds or *convolutions* separated by fissures. The outer part, or *cortex*, of the hemispheres is composed of gray matter, which consists mainly of ganglion cells and their interconnecting processes. There are numerous fibers which pass from the cortex to certain large ganglionic masses at the base of the brain and these in turn are connected with the spinal cord and with various cranial nerves. Then there are numerous fibers which run from one part of the cortex to the other so that the whole structure may be regarded as a great mass of nerve cells closely united by connecting fibers and intimately connected with other parts of the brain, with the spinal cord, and thence with other parts of the body.

Below the posterior part of the cerebral hemispheres is the *cerebellum*, an organ which probably has to do with the control or coördination of bodily movements, but of whose precise functions comparatively little is known.

The lowest part of the brain, the *bulb* or *medulla*, may be regarded as an enlargement of the spinal cord. From this part arise several pairs of nerves which are mainly distributed to the head, face and neck; one pair, however, the vagus nerve, sends branches to the lungs, heart and organs of digestion. This nerve has an important influence upon the beating of the heart and the movements of respiration. Destruction of a certain center in the medulla results in death, because respiratory movements are stopped and the animal dies of suffocation. An animal can live without a cerebrum, although he would be a very stupid sort of creature, but the medulla is absolutely essential to life.

When the cerebral hemispheres are taken out of a pigeon

the bird is at first dull and inactive, but after a time it regains its power of spontaneous movement. It can fly, avoid obstacles, balance itself on a perch, but it does not recognize its associates, pays no attention to its young and shows no evidence of any knowledge of the things it had previously learned. A German physiologist, Goltz, has succeeded in the difficult operation of removing the entire cerebrum from a dog, and in keeping the animal alive for several months. After recovery from the opera-

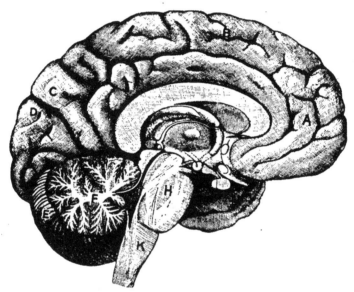

FIG. 214.—Median section through brain. *A, B, C, D,* lobes of the cerebrum; *E,* cerebellum; *F,* arbor vitæ; *H,* pons Varolii; *K,* medulla. (After Leidy.)

tion the dog was able to walk about spontaneously and even became unusually active. It would growl and snap when its paw was seized and it would eat food placed in contact with its nose, but it would reject meat or milk made bitter with quinine. While the dog could hear and could react to light, it recognized none of its old acquaintances either by sight or sound. Everything it had acquired by memory was lost, and it became a creature of pure in-

302   ANIMAL BIOLOGY

stinct, without knowledge and without understanding.

If intelligence is especially associated with the cerebral hemispheres can we say that different faculties of the mind are located in different regions of these parts of the brain? A number of years ago several theorists elaborated a system called phrenology, according to which various faculties of the mind are located in different parts of the brain, and therefore in order to determine what

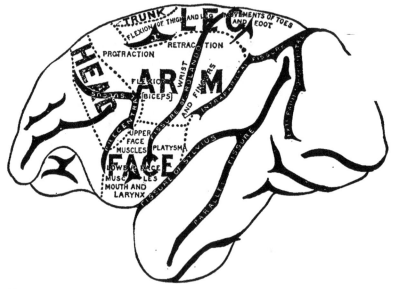

Fig. 215.—Left side of a monkey's brain showing motor areas. When these areas are stimulated movements are caused in the parts of the body designated by the names upon the corresponding parts of the brain.

faculties in a person were exceptionally well developed, it was only necessary to study the shape of his head which was supposed to give a fairly reliable index of the outline of the brain. Phrenology has now fallen into disrepute. Different parts of the cerebrum, however, have different functions, although various faculties of the mind cannot be mapped out as the phrenologists had them located. Near the middle of the cerebral hemispheres is the so-called motor area containing a number of centers which, when stimulated by electric needles, produce a movement in a particular part of the body. There is in the ape's brain, for instance, a center for the movement of the fingers, another for the forearm, another for the shoulder muscles, others again for various parts of the hind leg and trunk. When these centers are cut out the animal has difficulty in making movement in a corresponding part of its body. When brain tumors occur in the motor area their precise location is often indicated by the inability of the patient to perform certain movements. Many such cases have been cured by making an opening through the skull over the area indicated and removing the tumor.

Abnormal conditions of the brain are very frequently correlated with epilepsy and insanity. There is a remarkably close relationship between the activities of our brains and the working of our minds. It is not possible to have a healthy mind in a diseased brain, and as the brain is an especially delicate and sensitive organ it quickly feels the effect of injurious agencies. Nearly a fifth of the circulating blood goes to supply this organ. Nervous tissue absorbs a relatively large amount of material from the blood and it is the seat of active metabolic changes. This circumstance accounts in large part for the necessity for sleep which is a period of rest and restoration. During childhood, which is a time of rapid growth and active exercise, more sleep is required than later in life; but in all periods of life sleep cannot be dispensed with for long without producing very serious results.

CHAPTER XXXIV

THE ORGANS OF SENSE

We become aware of objects in the outer world through our organs of sense; these when stimulated set up impulses which are conveyed by sensory nerve fibers to the brain and arouse sensations of various kinds according to the kind of sense organ affected. Each sensation that we feel has its own peculiar sense organ which is especially sensitive to a particular activity in the outer world. Contact for instance stimulates the organs of touch, sound waves the organs of hearing, and light affects the organs of vision. Were it possible to destroy all of our sense organs or the nerves which lead from them to the central nervous system, the outer world would make no impression on us.

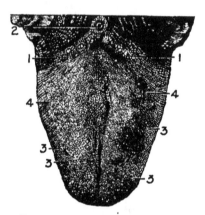

FIG. 216.—Tongue. 1, circumvallate papillæ; 2, circumvallate papilla, large; 3, toad-stool (*fungiform*) papillæ; 4, threadform (*filiform*) papillæ.

Some of our sense organs are distributed over most of the surface of the body, while others are limited to well-defined areas. We may feel sensations of touch over most of the skin and in various internal organs, but there are some areas such as the tip of the tongue and the ends of the fingers where tactile sensibility is especially acute. Heat and cold are likewise felt over most of the surface of the body. These sensations doubtless have their own separate nerves corresponding to little areas called heat spots and cold spots which, when stimulated, yield each its own sensation of heat or cold respectively. Menthol has the property of stimulating the cold spots and of making the skin feel cold, although in reality it may be quite warm. Pain is a definite sensation which is aroused when the tissues of the body are injured.

The sensation of taste is aroused by the stimulation of the taste buds of the tongue. There are a limited number

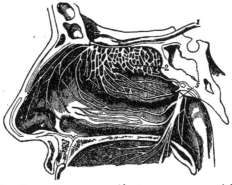

FIG. 217.—Nasal cavity. 1, olfactory nerve with its branches shown in 2; 3 and 4, turbinated bones; 5, fifth cranial nerve. (After Marshall.)

of true sensations of taste, *i.e.*, sweet, bitter, sour, salt; much of what we call tastes are really odors which are caused by substances arising through the pharynx and stimulating the olfactory nerves. This is why holding the nose when certain substances are being chewed and swallowed, and thereby preventing the air from freely entering the nasal cavity from behind, tends to make us unaware of their flavor.

The eyes are among the most complex and delicately adjusted organs of the body and they have long excited wonder and admiration on account of the perfection of their mechanism. The eyes are freely movable within their sockets, or orbits, by means of six small muscles. Their outer exposed surface is kept moist by the secretion of the *lachrymal glands*. Normally this secretion is drained

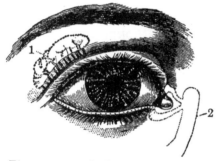

FIG. 218.—The eye. 1, lachrymal gland; 2, tear duct.

off by the tear duct which leads from the inner angle of the eye to the nasal cavity, but when the secretion is unusually abundant it may overflow as tears. The eye

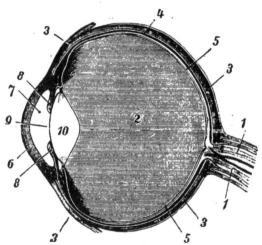

FIG. 219.—Sectiont hrough the eye. 1, optic nerve; 2, vitreous body; 3, sclerotic; 4, choroid; 5, retina, 6, cornea; 7, anterior chamber; 8, iris; 9, pupil; 10, crystalline lens.

is nearly spherical in form and is furnished with a tough, whitish, fibrous outer coat, the *sclerotic*, which extends over all but the anterior surface where it is replaced by the transparent *cornea*. Behind the cornea is a colored, circular partition, called the *iris*, which has a central, circular aperture, the *pupil*, through which light is admitted to the back part of the eye. Just behind the iris lies the transparent *crystalline lens*. This separates the interior of the eye into two chambers, (1) a small anterior one filled with a transparent fluid called the *aqueous humor*, and (2) a large posterior chamber filled with the transparent *vitreous humor*. The back part of the eye is lined internally by the sensitive *retina* which is really the expanded end of the optic nerve. Between the retina and the sclerotic is the black, pigmented *choroid coat* which serves to absorb scattered light which enters the eyeball.

The optical parts of the eye are so arranged as to throw images of outer objects on the retina. The eye has often been compared to a camera which, in a very similar way, is constructed so as to throw images on the photographic plate at the back. The lens of the eye functions like that of a camera in forming an image; the iris, which by contracting or relaxing alters the size of the pupil, corresponds to the shutter which regulates the amount of light entering the camera; the choroid, like the black inside of the camera, absorbs superfluous light; and the retina on which images are thrown is analogous to the sensitive photographic plate or film. As images formed by a lens are clearly outlined only when the object is a certain distance away, if the parts of the eye remained always the same we could not see both distant and near objects with equal clearness. We are enabled to do this because the curvature of the lens can be increased by the contraction of a special muscle that surrounds it, while the lens resumes its previous shape when the muscle is relaxed. The eye thus has the power of focusing itself upon objects at varying distances.

In many people the eye is not entirely spherical, or the lens has not the proper convexity, so that the eye does not have the usual range of vision. Some people are near sighted, that is, they see clearly only close at hand, while others are far sighted and can see clearly only

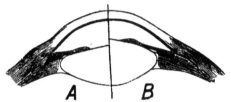

FIG. 220.—Diagram to illustrate the adjustment of the crystalline lens, A, for distant objects, B for near ones. (After Huxley.)

things that are distant. In the first case the clearest image is formed in front of the retina and in the other case the image is behind it. These defects are corrected by proper eye glasses.

A very common eye defect, called astigmatism, is caused by the unequal curvature of the eye ball or lens. This is

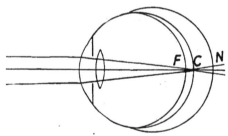

FIG. 221.—Diagram of the position of the retina. F, in far sight; N, in near sight; and C, in natural sight.

corrected by lenses which have a cylindrical instead of a spherical curvature. Much eye strain is endured by many people without their being aware of the fact; and many disorders, such as headache, nervousness, indigestion and other ailments are directly traceable to this cause. Reading in dim light or with the light in front so that we get

THE ORGANS OF SENSE 309

the glare from the paper tends to fatigue the eyes, and reading during convalescence from illness is especially apt to leave the eyes in a weakened condition.

The ear, or organ of hearing, is composed of three parts, the *external*, the *middle* and the *internal ear*. The external ears of most of the mammals can be turned so as to catch the sound, or else flattened against the head for protection, but in ourselves the ear muscles are mere rudiments capable of producing only a slight amount of movement,

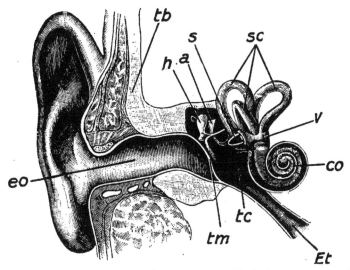

FIG. 222.—Section through the ear; *a*, anvil; *h*, hammer; *s*, stirrup; *co*, cochlea; *eo*, external opening; *Et*, Eustachian tube; *tb*, temporal bone; *tc*, tympanic cavity; *tm*, tympanic membrane; *v*, vestibule.

or none at all. In fact the aural appendages of man are almost devoid of function as they are poorly adapted for catching sound and represent but the vestiges of a structure once valuable to the animal ancestors of man. The passage leading toward the inner parts of the ear is furnished with numerous glands that secrete a waxy substance which serves to protect the ear from dust and insects. The inner end of the passage is closed by the *tympanic*, or *drum membrane*. On the inner side of this membrane lies the space called the middle ear. This is connected with the pharynx by the *Eustachian tube* which serves to admit air to the middle ear from the throat. When one swallows one can usually hear the opening of this tube.

Connected with the drum membrane on the one side and with the inner ear on the other is a chain of three small bones, called the *hammer*, the *anvil*, and the *stirrup*. Sound waves set the drum membrane into vibrations which are conveyed by these bones to the inner ear where they stimulate the end organs of the auditory nerve. The inner ear is a delicate, complex structure lodged in a cavity within the bones of the skull. The spirally wound *cochlea* which resembles a snail shell in shape contains the delicate end organs of the nerves of hearing. Another part of the internal ear consisting of three *semicircular canals* has a very different function, as it forms an organ for the maintenance of the equilibrium of the body.

CHAPTER XXXV

ALCOHOL AND TOBACCO

Human beings are not satisfied as a rule, with wholesome food and drink, but they manifest a perverse tendency toward the use of artificial stimulants and narcotics, whose influence on both mind and body is almost always injurious and frequently disastrous. While alcohol has been alluded to in discussing various bodily functions, its use is so deeply entrenched in popular custom that it is desirable to give here a fuller account of its physiological effects. The influence of alcohol has been studied by Professor Hodge by means of experiments on dogs, extending over five years. Four pups of the same litter were chosen and observed for some time to see if they showed any differences previous to giving them alcohol. The two most vigorous pups were given a daily portion of alcohol, but never enough to produce intoxication. In a few months the alcoholic dogs became more sleepy and lifeless, whereas the other dogs which were given no alcohol were bright and active. Experiments in which all four dogs were given a hundred trials in chasing a ball thrown to a distance of a hundred feet showed that the alcoholic dogs secured the ball only about one-half as often as the others. The alcoholic dogs were much more nervous, and when all the dogs contracted the distemper, the disease was much more severe on those that were given alcohol.

These results are quite typical of the effects of alcohol on human beings. Dr. Parkes had the opportunity of testing the influence of alcohol on two lots of soldiers. The men were of the same age, had the same food, and lived under the same conditions. The one lot of men were given beer when they wanted it, which they usually did when tired; the others had no alcohol. While the alcoholic lot outstripped the others at first, they soon lagged behind and did far less in a day than the others. When the conditions were reversed, the lot of men who formerly drank but now took no alcohol did much more work in a day than the other lot. The experiences of Kitchener, Roberts, and others with soldiers has convinced them that men endure marching and other tasks much better without alcohol. Count von Haeseler, the German commander, says "The soldier who abstains altogether is the best soldier. He can accomplish more, can march better, and is a better soldier than the man who drinks even moderately."

The almost unanimous testimony of mountain climbers is that if arduous journeys are to be taken no alcohol should be used. Athletes in training are usually not allowed alcohol. Although alcohol is a food, the trainers of athletes have learned that it is a very dangerous experiment to allow its use, even in very small amounts.

Alcohol not only reduces the capacity to perform tasks involving strength and endurance, but it has an even greater effect on performances that require dexterity and skill. Experiments with type setters who were given a moderate amount of alcohol on certain days and no alcohol on others, have shown that the amount of work done when no alcohol is given was, on the average, markedly less on the alcoholic days, although the men were under the impression that they were accomplishing more. Kraepelin, who believed that alcohol in small amounts increased the activity of his mind in adding, subtracting, and learning figures, found, when he came to test the matter, that he accomplished these operations less quickly than when he had taken no alcohol. Numerous experiments with Swedish soldiers under various conditions have shown that accuracy in shooting at a target was reduced from 30 to 50 per cent. when a small amount of brandy was given.

In the higher operations of the mind alcohol acts as a depressant and inhibitor. A person slightly under its influence often talks more rapidly and appears more lively, but he acts with less judgment. Herbert Spencer observes "Incipient intoxication, the feeling of being jolly, shows itself in a failure to form involved and abstract ideas." Helmholtz, in speaking of the inspirations that came to him while pondering over his problems, said "They were especially inclined to appear to me while indulging in a quiet walk in the sunshine or over the forest-clad mountains, but the smallest quantity of an alcoholic beverage seemed to frighten these ideas away."

Alcohol in inhibiting the higher operations of the mind causes the loss of self-restraint that only too frequently leads to crime. Swedish statistics show that out of 24,298 prisoners committed to hard labor, 17,374, or 71.2 per cent. attributed their crime to the use of alcohol. Dr. Sullivan found that "out of 200 men convicted of murder or attempts at murder, 158 were of alcoholic habits, and in 120 of these, or 50 per cent. of the whole series, the criminal act was directly due to alcoholism." The same inhibition of higher nerve centers accounts for the large percentage of accidents that happen to people influenced by drink. The nervous system is a delicate mechanism, and when anything important depends on its proper working, alcohol had better be left entirely alone.

While the evil effects of over indulgence in alcoholic drinks are evident enough, it is often claimed that moderate drinking can be practised with no serious results.

Many people, it is true, drink regularly a small amount of alcoholic beverage and live to an advanced age in apparent good health. But this does not prove that moderate drinking was not harmful to them, or that in other people less able to withstand the strain it may not cause greater injury. One of the worst effects of moderate drinking lies in the danger that it may lead to heavy drinking. Many people, once they acquire a taste for alcohol, soon become victims of a habit which is insidiously fastened upon them and finally makes a wreck of their lives. That many men of fine intellect and excellent character become the slaves of alcohol is a fact known to everyone, and whoever takes to drink thinking that it will be easy to break off is performing a very dangerous experiment. Habit-forming drugs impair the will at the same time that they increase desire. And before one is aware of the danger, he may be already in the clutches of the enemy that he is powerless to shake off.

One of the best evidences of the injurious influence of moderate drinking is afforded by the statistics of life insurance companies which show that the average length of life of moderate drinkers is less than that of total abstainers. Some life insurance companies do not take even moderate drinkers. In one of these, the American Temperance Life Insurance Association, the death rate of the members is 26 per cent. less than that of general risk. Users of alcohol are more liable to contract infections such as pneumonia, tuberculosis, and cholera, and the disease is, as a rule, more severe with them.

Tobacco

Ever since its first introduction into the civilized world by Sir Walter Raleigh, tobacco has been very extensively used by a considerable part of mankind. Tobacco has a narcotic effect which is due to small quantities of nicotine, a substance so poisonous that only two or three drops are required to cause the death of a man. Habitual users are affected but little by this small percentage of nicotine, but those just beginning the use of tobacco often have a very uncomfortable time after their first chew or smoke.

Tobacco, like other habit-forming drugs, creates a craving which is frequently difficult to overcome and which may lead to an immoderate use that is decidedly injurious to health. Many people, however, may use moderate amounts of tobacco for years with no noticeable ill effects. In such cases one has to reckon with the possibility of less obvious injury which ordinarily escapes detection. Where men are in training for an athletic contest which requires all their muscular and nervous energies, it is found by athletic trainers generally that it is best to forbid all use of tobacco.

The effect of tobacco on the heart in producing the condition known as "tobacco heart" is well known. Tobacco smoke is frequently irritating to the throat, and also to the lungs, especially when it is inhaled, as it often is, by smokers of cigarettes.

All students of the subject agree that tobacco exerts a very harmful influence upon young boys. It stunts their growth, saps their vitality and dulls their intellect; a school boy who is addicted to its use is almost sure to make a relatively poor record. Even among college students the use of tobacco only too frequently goes along with idleness and poor scholarship. Andrew D. White, former President of Cornell University, remarked that "I never knew a student to smoke cigarettes who did not disappoint expectations;" and Dr. Meylan of Columbia University, who has made a careful study of smokers and non-smokers among the students, states that "the scholarship standing of smokers was distinctly lower than that of the non-smokers."

There is no reasonable doubt that people would be better off without tobacco than with it. The tobacco habit is expensive, often repugnant to others, decidedly harmful to youth, and frequently injurious to adults. Many derive solace from their pipe or cigar. They may think that whatever bad effect tobacco may have upon them is more than outweighed by the satisfaction derived from its use. If there are people for whom this is true—and it is by no means certain that there are—it is perfectly clear that it is true only for those who have reached maturity. As Peabody has remarked, "the ambitious boy who has any regard for developing a vigorous body fitted for athletic success, for training a mind capable of clear thinking, and for preparing himself for a successful life work, will resist all temptations to smoke, at least until he has attained his full growth."

CHAPTER XXXVI

BACTERIA AND DISEASE

For ages mankind has been nearly helpless before the ravages of contagious diseases. Ignorant of their real cause, people have attributed such diseases to "effluvia," to poisons carried in the air, and in former ages to possession by evil spirits. We know now that they are produced by some kind of an organism that can be transferred from one individual to another. Many diseases are caused by protozoans, as is the case with malaria, Texas fever and amœbic dysentery. Some diseases (ringworm) are produced by fungi. A few diseases, such as filariasis, trichinosis and the itch, are caused by higher animals. But the greater part of our contagious maladies are due to very minute organisms known as bacteria.

These bacteria are the simplest known forms of life. They are as a rule exceedingly minute. A great many species appear like short rods (Bacilli), some are nearly spherical (Cocci), while others are spiral (Spirillum, Spirochetes). Some of the species are furnished with one or more flagella by means of which they may move about. Bacteria usually multiply by fission, and so rapid is their multiplication that a single bacillus may give rise to 16,700,000 individuals in twenty-four hours. At times, and especially under unfavorable conditions, bacteria may produce small, rounded bodies called spores which are unusually resistant to heat, cold or dryness. When better conditions occur these spores give rise to other bacteria.

There are countless different kinds of bacteria; they live under the most diverse conditions, and are capable of subsisting upon a great many kinds of food. Many forms cause the decay of the bodies of higher organisms. Whenever an organic body putrefies or decays, it may be found to be teeming with multitudes of bacteria. If these minute organisms are excluded from an organic body it may be preserved for a long period. Canning fruits, vegetables and meats is essentially a device for keeping these articles free from bacteria. The heating of canned goods kills whatever bacteria may have been present in them, and the sealing prevents the access of others. Once introduce bacteria and the substances soon decay.

Many articles are kept from spoiling by means of preservatives, or substances which either kill bacteria or check their growth. Up to a point which varies with different species, bacteria multiply more rapidly as the temperature is increased, and they are kept from multiplication at a temperature at or near the freezing point. Cold storage therefore keeps them from attacking meats and other articles of food; hence our cold storage plants, refrigerator cars and ice chests.

Bacteria are usually not killed by freezing, however, and many kinds will endure a temperature of over $100°C$ below zero without losing their vitality. Ice may harbor the germs of many diseases and in particular those of typhoid fever which are especially hard to kill. Boiling kills most bacteria, but the spores of many species will resist even boiling for a certain period.

Bacteria are almost universally distributed in water, soil and all sorts of organic material. Owing to their minute size they are capable of being easily carried through the air. If a bit of beef broth or vegetable infusion is left exposed to the air for a few moments it will become infected with bacteria; in fact they often gain access when it is thought that they are effectively excluded. They are especially liable to be carried about on particles of dust. Look at a beam of light entering a room and you will usually see myriads of small bodies floating in the air. For every particle that you can see there are thousands that are too small to be visible. When we remember that even invisible particles may be over a million times as large as a bacillus we can gain some idea of what it means to effectively exclude the possibility of bacterial infection.

While many bacteria are harmful, most species are probably beneficial. The rôle of bacteria in causing decay is one of great importance. Organisms are resolved back into their primitive constituents, and their atoms may live again in the bodies of other forms of life. On account of their causing the decomposition of organic material the bacteria play an important part in the production of soil. Certain species associated with the roots of plants have the property of utilizing the nitrogen of the air and converting it into a form that may be subsequently used by plants and animals. Many of the processes of fermentation, such as the souring of milk and the production of vinegar, are the result of bacterial activity. In fact were it not for these invisible organisms higher forms of life would not be able to exist in the earth.

That certain species of bacteria should come to live in the tissues of plants and animals is quite analogous to the fact that many other organisms have adopted a parasitic mode of life. The disease-producing bacteria do not wait until an organism is dead before they attack it, but like the bacteria that effect the decomposition of the dead body they tend to bring about the dissolution of the organism. Louis Pasteur, the great French investigator who did more than anyone else to establish the germ theory of disease, was led to his most important discoveries on account of his previous studies of the phenomena of fermentation and putrefaction. Once the germ theory of disease was established, contagion was no longer a mystery, but a natural result of the transfer of minute organisms from one person to another. Just as a drop of decaying substance will set up decomposition in fresh material, so will a small amount of matter from a diseased person convey the disease to a healthy individual. It has been shown that, in a great many diseases, bacteria are uniformly present in great numbers in the tissues of the person affected. In many cases it has been found possible to cultivate the germs of certain diseases in artificial media outside the body, and to propagate them free from admixture with other germs. The germs from such "pure cultures," as they are called, have been shown to give rise to the disease in question when inoculated into the body of a healthy person.

When the secret of contagion was known, it became much easier to check the spread of contagious diseases. For this purpose much use is made of germicides and disinfectants, substances which kill the germs of the disease. Rooms that have been occupied by diseased persons are commonly fumigated with formaldehyde, a strong germicide, before they are again occupied. Articles used in connection with the patient are washed in an antiseptic solution or boiled. And the patient is so far as possible kept free from contact with healthy people until disease germs are no longer given off from his body.

Formerly surgical operations were commonly attended with gangrene, blood poisoning and other infections which we now know are caused by bacteria. Surgeons now exercise the greatest care in keeping the wounds of their patients free from bacterial infection. All instruments used are carefully sterilized and everything connected with the operation is made scrupulously clean. As a result operations are now performed which were undreamed of before the days of aseptic and antiseptic surgery, and the number of infections following ordinary operations has been greatly diminished.

Cuts, scratches and abrasions of the skin, while they usually heal up with no serious results, may become infected and lead to blood poisoning. It is well to wash them, therefore, with peroxide of hydrogen or some other antiseptic, and then bind them up so as to exclude the entrance of other germs. Boils and carbuncles result from bacterial infection and the pus they contain may give rise to similar infections in other parts of the body, or in the body of another person. The common notion that boils are useful in eliminating impurities from the blood is absurd. On the contrary they are a source of actual poisoning to the whole body.

The number of diseases caused by bacteria and protozoa is very great. We shall describe, therefore, only a few of those about which everyone should have some knowledge for the sake of his own safety.

Colds.—What we commonly call colds are really infections. Almost everyone has noticed how colds tend to run through a family or a school, and how at times colds are unusually prevalent. People commonly believe that they "catch cold" by sitting in a draught, getting their feet wet, or exposing themselves in cold weather. What really happens is that their temperature or their resistance may be reduced by these circumstances, and thus an opportunity is offered for the germs of the infection to make headway, whereas otherwise they might have been kept in abeyance. Arctic travellers and people living away from contact with their fellows rarely take cold despite their exposure. On the other hand, colds are often contracted when there has been no exposure to cold at all.

A little attention to the subject will bring home to us how many are the ways in which the germs of colds and other diseases may be transferred from one person to another. The afflicted person may cough or sneeze and send into the air a multitude of germs which may be breathed in by other people. Most infections, however, are probably carried by the hand-to-mouth method. A person with a cold for instance, by the frequent use of a handkerchief, inevitably transfers the germs that abound

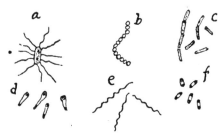

FIG. 223.—Forms of disease-producing bacteria. *a*, typhoid; *b*, staphylococcus; *c*, tuberculosis; *d*, diphtheria; *e*, syphilis; *f*, plague.

in his mucous secretions to his hands. He may shake hands with another person or handle some object that the other person touches. The latter may eat some food that he handles or otherwise bring the germs on his hands in contact with his mouth and thus the transfer is made. Door knobs, straps in street cars, or any objects commonly handled by several people afford excellent means for the transfer of the germs of disease. If one has been exposed to any of these sources of contagion, especially during a time when colds or other infections are prevalent, care should be taken not to eat before the hands are thoroughly washed. It is well to avoid people with colds whenever possible, and the sufferer should bear in mind the possibility of his infecting others and give as little opportunity as he can for the transfer of his affliction. The term cold is loosely applied to a number of infections which vary greatly in their severity. Colds often pave the way for tuberculosis, pneumonia and other ills, and they should be treated with promptness and vigor when first contracted, otherwise they are much more difficult to deal with.

Tuberculosis, one of the most frequent causes of death, is a disease that may attack any part of the body, although it most commonly affects the lungs when it is known as consumption. It was proven by Robert Koch, a famous German bacteriologist, to be caused by a small rod-like organism, the *Bacillus tuberculosis*. Cattle are commonly affected with a form of tuberculosis and the bacilli frequently occur in milk. Much effort has been made to stamp out the disease in cattle, and milk that is sold in cities is often put through a process called pasteurization in order to kill the germs of tuberculosis or other disease germs that may be contained in it. It is now recognized that tuberculosis may be conveyed from one person to another. This may be done by the hand-to-mouth method previously described, but a frequent source of infection consists in breathing air laden with tubercle bacilli. Air in dwellings occupied by consumptives is particularly dangerous, especially if it contains much dust. The sputum of consumptives usually abounds in bacilli, and when dried and reduced to powder it may be readily blown about in the air and taken into the lungs. Consumptives should be particularly careful not to expectorate in places where their dried sputum can possibly be a source of infection. The sputum should be disposed of in receptacles especially devised for the use of consumptives. Care should be taken, when coughing or sneezing, to cover the mouth or nose with a handkerchief, and the dishes used by consumptives should be carefully sterilized by boiling.

The isolation of tuberculous patients and the care that has been taken to prevent the dissemination of the germs of the disease has resulted in a marked diminution of tuberculosis in recent years. With the spread of knowledge regarding the transfer of tuberculosis, and with continued efforts to prevent its spread, there is reason to believe that this "great white plague" will finally be exterminated.

Tuberculosis is usually curable if treated in the early stages. Rest, plenty of nutritious food, such as fresh eggs and milk, and living continually in the open air will, in the majority of cases, effect a cure.

Typhoid fever is a disease that is carried mainly through food and drink. The bacillus that causes it is unusually resistant to heat and cold, and may live in the water and in soil and sewage for a long period. The germs are especially abundant in the intestine and hence in the excreta of typhoid patients. Where sewage is allowed to discharge into rivers or lakes, the germs are liable to be taken in in drinking water, and many epidemics of typhoid in cities have been traced to the contamination of the water supply. Sewage from one town is often allowed to flow into a stream that forms the water supply of another town farther down. Many cities profiting by costly experience, have instituted filtering plants for purifying their water. Some cities situated on lakes have elaborate plants for disposing of their sewage instead of pouring it into the water that they use for drinking. Country places which derive their water from wells located so that the seepage from privies may flow into them are often afflicted with typhoid. In all these cases the elimination of typhoid depends upon the purification of the water supply.

Germs of typhoid may be carried in food. Several epidemics have been traced to oysters grown in places contaminated by sewage. In many instances the disease has been carried by flies which alight upon food after having walked over excreta.

People who have apparently recovered from typhoid are sometimes capable of spreading the disease for years

FIG. 224.—Pollution of a well by the contents of a neighboring cess pool.

afterward. These people who are called "typhoid carriers" harbor multitudes of typhoid bacilli in their intestines and are especially dangerous if they handle food that is to be eaten by others. The case of "Typhoid Mary," who served as a domestic in several homes and left a trail of typhoid patients wherever she went, is one of the best known.

Most diseases tend to run a certain course and end in natural recovery. If the disease does not prove fatal, the body conquers the disease and very frequently one attack protects the individual from another attack of the same malady. Typhoid, scarlet fever, smallpox, are rarely taken more than once, the patient having acquired what is called *immunity* to these diseases.

Disease germs produce their deleterious effects by generating some poisonous substance, or *toxin*. The cells of the body have the property of producing substances which neutralize or destroy these poisons, and these are called *antitoxins*. The white phagocytes of the blood also attack and devour disease-producing bacteria. A disease in the body means a battle between the bacteria, which tend to live and multiply at the expense of the organism, and the cells of the body with their antitoxins and phagocytes. The discovery that bacterial poisons may be destroyed by antitoxins generated by the body has led to efforts to control diseases by injecting antitoxins into the blood. One of the first and most noteworthy attempts of this kind is the antitoxin treatment of diphtheria. This disease formerly had a high death rate. The rod-like bacilli occur chiefly in the throat and generate a toxin of extreme virulence. The German bacteriologist, von Behring, found in 1892 that if diphtheria toxin is injected into a horse, the blood serum of the animal will contain a substance that neutralizes the toxin. In the manufacture of diphtheria antitoxin healthy horses are given several injections of diphtheria toxins of gradually increasing strength. After several months some of the animal's blood is removed and the serum preserved for injecting into human beings. The antitoxin treatment of diphtheria is now regularly employed and it has reduced the death rate of this dreaded disease by 75 to 80 per cent.

Rabies, or hydrophobia, has been successfully treated by a method somewhat different from the preceding, but based on the principle of rendering the body immune to the disease. This disease occurs on dogs and may be communicated by biting to other kinds of animals and to man. Remedies were formerly powerless against this disease; unless the germs were killed by promptly cauterizing the wound, the patient had no hope of escaping one of the most horrible kinds of death. To the genius of Louis Pasteur the world owes the discovery of a method of cure now known as the Pasteur treatment. By a series of injections of a preparation made from the spinal cord of a mad dog the patient may usually be prevented from contracting hydrophobia, even if the treatment is begun several days after the bite. A very high percentage of those bitten by rabid dogs contract hydrophobia. The Pasteur Institute at Paris has treated many thousands of such cases with an average mortality of less than one-half of 1 per cent. Since the recent outbreak of rabies in California, 641 persons have been treated by virus supplied by the Hygienic Laboratory at Berkeley, Calif. According to Dr. Geiger, "Eliminating all persons treated who were not bitten, the percentage of failures with virus supplied by this Bureau was 0.491, less than one-half of 1 per cent." In over 98 per cent. of the persons bitten, the animals doing the biting were found by laboratory examination to have had rabies.

There is no foundation for the belief that dogs are caused to go mad by the hot weather of "dog days." All suspected dogs should be confined until the time for symptoms of hydrophobia to appear are past. By muzzling all dogs whenever there is the least danger of hydrophobia this disease could soon be stamped out, but through carelessness that is inexcusable in the light of our present knowledge, hydrophobia in many localities has actually been on the increase in recent years.

Another method of securing immunity to disease is by the introduction of the germs of a milder form of the malady. This is the procedure followed in vaccination to prevent smallpox. In 1796 the English physician, Edward Jenner, found that virus taken from cattle with the so-called cow pox would, when inoculated into the human body, prevent the individual from taking smallpox. A properly vaccinated person, if exposed to smallpox, may contract a much milder malady called varioloid, but vaccination decreases his chances of doing even this. Since vaccination has become prevalent smallpox has very greatly decreased. In the 18th century it is estimated that 18,000,000 died of smallpox. In Russia alone from 1893–1897 there were 275,502 deaths from this disease, while in Germany which had compulsory vaccination there were no epidemics. The German army has had but two deaths from smallpox since 1874. Before vaccination was introduced into Sweden the death rate from smallpox was 165 per 100,000, but since vaccination was made compulsory it fell to 18 per 100,000. Formerly undesirable after effects were sometimes produced by vaccination, but with improved methods of securing pure virus the dangers are now very slight. A form of vaccination is now used with remarkable success in typhoid fever. Small amounts of the toxin obtained from dead typhoid bacilli are introduced at different times and the person so treated is rendered comparatively immune from typhoid for a considerable period. In the United States army since vaccination against typhoid was required the death rate from typhoid has been less than one-fiftieth of what it was before. The death rate from typhoid during the Mexican war and the civil war was appalling. In the Spanish war "5000 men in the United States army died of typhoid or other fly-borne diseases while only 300 were killed by Spanish bullets."

Since contagious diseases constitute so serious a menace to human welfare it is necessary to have laws to regulate the treatment of contagious cases, so that they may be prevented from infecting others. People with such diseases are often put in quarantine until danger of spreading their infection is past. Immigrants are examined for contagious diseases and detained for a certain period before being allowed to land. Through these means many epidemics have doubtless been averted which otherwise would have destroyed thousands of lives. In cities and towns boards of health and health officers look after the enforcement of regulations for checking disease, and for securing general sanitary conditions.

The correct diagnosis and treatment of disease demands expert knowledge and special training, and states have wisely framed laws requiring that those who occupy themselves with the very responsible business of the physician or surgeon should have received adequate instruction for carrying on their work. Even the best physicians sometimes make mistakes, but the ignorant or careless charlatan does little but harm. Requirements for obtaining a license to practice medicine are steadily being raised but there are still many incompetents in the regular profession and many quacks who contrive to evade the law and carry on a lucrative trade. One of the favorite devices of the quack is advertisement. Newspapers commonly contain several advertisements of the wonderful skill and success of various doctors who often pretend to cure ills for which no remedy has ever been discovered. It is a safe rule to avoid all so-called doctors who advertise, for the great majority of them are unprincipled charlatans.

Numerous frauds are imposed on the public through

patent medicines. Some of these preparations are good, but many others which have been widely advertised are worthless, if not positively injurious. The soothing syrups for quieting crying children are among the worst of these, because they nearly always contain opium or some other drug which works lasting injury on the helpless child.

PART III
GENERAL FEATURES AND ADAPTATIONS

CHAPTER XXXVII

THE PERPETUATION OF LIFE

All organisms have the property of producing other organisms similar to themselves and thus continuing their race. In the simplest forms of life new individuals commonly arise by the division or fission of the parent form. An Amœba or Paramœcium, as we have seen, simply constricts in two and the new individuals soon regain their

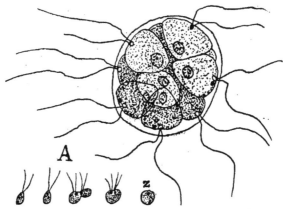

FIG. 225.—*Pandorina morum.* *A*, gametes formed by repeated divisions of the cells of the colony. These gametes meet and fuse and finally lose their flagella and become converted into a spherical encysted zygote, *z*.

normal size and shape. Other organisms reproduce by budding, such as most hydroids, sponges, several kinds of worms and many other primitive animals. In some of the Protozoa the body divides up into a number of bodies called *spores* which scatter and develop new individuals.

In organisms except the very simplest, such as the bacteria, the process of reproduction is commonly associated with the phenomenon of sex. We have already described some of the manifestations of sex in the conjugation of Paramœcium where after a series of generations produced by fission there is a union of individuals by pairs, during which each individual receives a nucleus from the other.

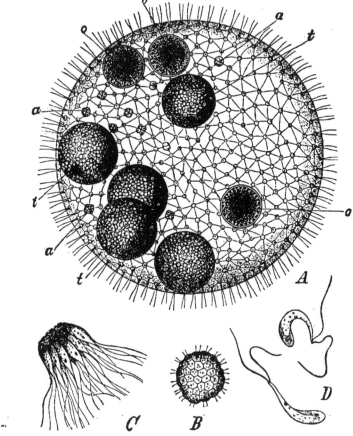

FIG. 226.—*Volvox aureus.* (After Klein and Schenck.) *A*, mature colony containing daughter colonies; (*t*) and ova (*o*); *B*, group of 32 developing spermatozoa seen end on; *C*, the same seen sideways; *D*, mature spermatozoa, × 824.

After the union of one of its own nuclei with the one received from its partner, the individuals separate and continue to divide by fission as before. More commonly in primitive organisms there is a complete fusion of nucleus and cytoplasm so that each conjugating individual loses its identity in the resulting product. In some of the one-celled animals and plants the conjugating individuals, instead of being of equal size, have become differentiated into larger, relatively passive individuals on the one hand, and smaller, more active ones on the other. In the colonial flagellate Pandorina the conjugating cells are both active and but slightly unequal in size, but in the related form Eudorina and in Volvox the conjugating cells are very dissimilar. One kind is large, spherical in form and

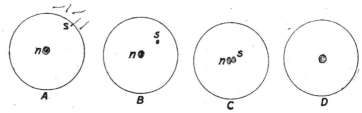

FIG. 227.—Fertilization of the egg. *A*, egg entered by a sperm cell, *s*; *B*, the nucleus of the head of the sperm cell, *s*, enlarged and approaching the nucleus of the egg, *n*; *C*, sperm nucleus, *s* more enlarged and in contact with the egg nucleus *n*; *D*, egg nucleus and sperm nucleus fused together.

devoid of activity; while the other is small, with an ovoid head and a lash-like tail by means of which it swims through the water. One of the small active cells meets and fuses with, or fertilizes, one of the large ones which, after going into a resting stage, produces by repeated division a new colony.

In all the multicellular animals the sex cells are differentiated into two very sharply contrasted types, the relatively large and inactive *ova*, or *egg cells*, and the small active *spermatozoa*, or *sperm cells* whose function it is to meet and fertilize the eggs. The eggs of nearly all animals require to be fertilized before they can develop, but in exceptional cases eggs may develop without being fertilized. This process which is called *parthenogenesis* (see page 35) is well illustrated by the generation of the aphids or plant lice in which the females may reproduce by parthenogenesis for several generations, especially during the summer. After a time, however, different kinds of eggs are produced which require fertilization before they develop. It has been found by Loeb that in some animals eggs which normally require fertilization before developing, may be stimulated artificially by chemicals and other agencies so that they develop without fertilization into apparently normal embryos. In almost all forms in which parthenogenesis occurs, reproduction by means of fertilized eggs occurs also after one or more parthenogenetic generations.

While in most animals sperm and egg cells are produced by separate male and female individuals, there are many animals in which both kinds of sex cells are borne in the same body. Such animals are called *hermaphrodites*. Most flat worms, earthworms, leeches, land snails and tunicates are hermaphrodites, and there are occasional hermaphroditic species in many other groups of animals, and exceptional hermaphrodites which arise as "sports" or monstrosities in species with normally separate sexes. It is a curious fact that in hermaphroditic animals the eggs are very rarely fertilized by sperms from the same individual, but instead there is cross fertilization, as we have seen in our account of the earthworm. In sexual reproduction in general there is a mingling of germinal material derived from two separate individuals.

After the union of the egg and sperm cell there begins the process of embryonic development which results in the formation of a new individual. This process, which is one of great complexity, forms the subject matter of the science of embryology, a subject which can be treated but very briefly in an elementary book. The egg in all animals consists of a single cell. Usually this is of minute size, but in birds and many reptiles and in a few other forms it may be very large. In the egg of a bird the original egg cell forms the part which we call the yolk. This yolk after its discharge from the ovary receives first a coating of albumen, or white, and then the shell during its passage down the oviduct, so that the egg in this case consists of an enormous cell, the yolk, plus the surrounding materials which are secreted by the glands of the oviduct.

Animals which lay eggs are said to be oviparous, but there are many animals such as all the mammals (with the exception of the monotremes) in which the development of the embryo takes place in the body of the mother. These forms are called *viviparous* since they bring forth living young, but in all viviparous animals the embryo arises from an ovum or egg, just as in those forms which lay eggs. When eggs are of large size it is due to the presence of yolk or other material which affords food for the developing embryo. The eggs of the viviparous mammals are very minute. They are always fertilized within the body of the female, usually while in the oviduct, and they undergo development within the uterus. In all mammals higher than the marsupials the embryo becomes attached to the wall of the uterus by an organ containing numerous blood vessels, called the *placenta*. Through this organ nutriment is carried by means of the maternal blood to the embryo, whose own blood vessels extend into the placenta and there absorb the food material that passes out by osmosis from the blood of the mother. The blood vessels that lead from the embryo to the placenta are contained in the umbilical cord which enters the embryonic body at the point called the navel. The placenta is cast out soon after birth.

The first clearly marked stage of development that occurs after the fertilization of the egg is the process of *cleavage* by which the egg is divided into a number of cells. Typically cleavage results in the formation of a hollow sphere of cells, the *blastula;* this becomes pushed in on one side forming a double-layered sac called the *gastrula* whose inner wall generally gives rise to the lining of the digestive cavity and the various organs that arise from it. In some animals, especially those in which the egg contains much yolk, the blastula and gastrula stages may become very much modified, so that they are

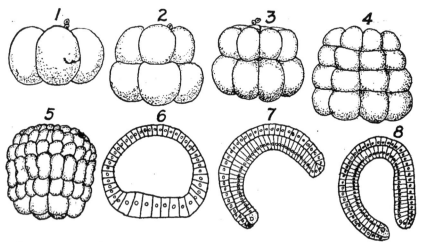

Fig. 228.—Early development of Amphioxus. 1, four cell stage; 2, eight cells; 3, 16 cells; 4, 32 cells; 5, blastula; 6, blastula cut through the middle; 7, early, and 8, late gastrula. (From Hatschek.)

not easily recognized, but in most of the multicellular animals they are nevertheless represented in some form. In the gastrula stage there are formed the two primary germ layers, an inner, the *entoderm*, and an outer, the *ectoderm*. In animals above the cœlenterates a third germ layer, the *mesoderm*, arises between the two others. At first the cells of the embryo are similar in kind, but, as the organs develop from the germ layers, the cells become gradually differentiated into all the varied kinds found in the adult body.

In the early stages of development animals in general are much more alike than they are later. As they develop the embryos of different groups become more and more dissimilar; those of distantly related groups are strikingly different in their early stages, while the embryos of closely related forms usually show a strong resemblance in all periods of development. The embryos of higher animals are often similar to the adult stages of animals which stand below them in the scale of life. These similarities, as we shall see in a later section, point clearly to a descent of the higher animals from more primitive forms.

Some animals when hatched or born bear a fairly close resemblance to the adult condition, but in many others the young as they first emerge are so different from their parents as to appear to belong to a quite different group of animals. Such animals are said to undergo a *metamorphosis* in developing into the mature form. Instances are furnished by the butterflies and moths, beetles, flies and many other insects, by the transformation of tadpoles into frogs, and by the larval stages of many marine invertebrates. In several cases larvæ have been described as new types of life before it was discovered that they represented the young stage of some previously known animal.

In higher animals a part of the function of perpetuating life consists in caring for offspring until they are able to shift for themselves. Among low forms the young receive no attention whatever from their parents, most of whom do not recognize their offspring as their own. The young spider is a nimble, active creature, that can spin its own web practically as well as in later life. Many Crustacea carry their eggs and for a time their young in brood pouches or attached to appendages of the abdomen, but they never care for their offspring in any way, and are quite as willing to devour them as any other kind of food. The young of most insects do not receive or need any attention from their parents, although in some of the social Hymenoptera the young are fed and tended with scrupulous care.

FIG. 229.—*Dolomedes mirabilis* carrying her cocoon. (After Blackwall.)

Among fishes the eggs, after they are shed and fertilized, are usually left to their fate; but some forms such as the stickleback and the dog-fish Amia build nests for the eggs which are watched and defended, usually by the male who protects them from enemies, but parental solicitude lasts for only a short time.

The amphibia and reptiles show little concern for their offspring, but in the birds and mammals we find parental care well developed especially in higher forms. The primi-

FIG. 230.—Nestling marsh hawks. (After Baker.)

tive birds as a rule construct crude nests and the young are hatched so that they are able to look out for themselves, either at once or at a very early period. Among the higher song birds the nests are more carefully made.

The young which are hatched in a weak and helpless condition are fed, brooded and protected by their parents who usually keep the nest clean and often continue to care for their young after they have left the nest and are apparently quite able to secure food for themselves. Among primitive types which produce a large number of offspring there is a great waste of life. A codfish may lay over 9,000,000 eggs of which as a rule only two produce fishes

Fig. 231.—Bluebird at edge of nest with grasshopper in mouth for young. (After Baker.)

that live to a mature age. A humming bird which carefully rears its young in a well-prepared nest lays but two eggs in a season. With parental care a species no longer needs the enormous fecundity of the primitive forms which leave their eggs and young to the mercies of the elements and numerous enemies.

In the mammals parental care is universal. The young are closely dependent on their parents for food which is supplied by the mammary glands and the instinct to suckle and protect the young is a part of the endowment of every mother. When the need for milk and protection is past parental affection as a rule soon passes into indifference. The period of relatively helpless infancy increases as we pass to higher mammals. This affords a greater opportunity to learn by experience while under the protection and guidance of parents before the young animal has to face the serious business of life. Young birds are taught to fear particular enemies and to peck at certain kinds of food. The danger chirr of the old hen fills the young chick with alarm and the alarm notes of the partridge will send the young into hiding places in the grass. The lessons learned in infancy frequently are the means of saving life when the young are free from parental guidance.

Among the simplest organisms, as we have seen, the perpetuation of life is effected simply by the process of fission. In somewhat higher forms we meet with phenomenon of sex, and the various activities of mating. Later we find that the activities of reproduction involve the care of eggs and young; and as we ascend the scale of life the time and energy expended upon the rearing of offspring becomes greater and greater. The perpetuation of the race finally comes to mean not only the production of new individuals, but the fostering and training of the new generation until it is capable of leading an independent life.

CHAPTER XXVIII

THE EVOLUTION OF LIFE

The brief survey of the animal kingdom to which the first part of this book is devoted gives some idea, inadequate though it be, of the variety of animal life on the surface of the earth. Zoologists have described several hundred thousand species, and the fact that new species are being described at the rate of about ten thousand a year shows us how far we still are from having a complete list of the earth's fauna. The number of species of animals now on the earth is certainly over a million and is possibly several millions. But the number of species now living constitute but a small fraction of the enormous number that formerly peopled the earth. The science of Geology teaches us that the crust of the earth is a great burial ground in which are interred the remains of countless animals and plants, and that new forms have constantly replaced the old during the many millions of years involved in geological history.

The question naturally arises: How did all this wealth of plant and animal life come into existence? Formerly it was generally held that each species was separately created, but as students of life came to have more extensive knowledge of the structure, distribution and relationships of living forms, and as they traced the succession of extinct species buried in the rocky strata of the earth's crust, they became convinced, almost without exception, that species of plants and animals arose by a gradual process of development or evolution. How life first began no one knows, but the gap between the non-living and the living once having been bridged, life gradually advanced from the simplest particles of living substance to the highest types of plants and animals. The establishment of the doctrine of evolution was one of the greatest of the scientific achieve-

FIG. 232.—Charles Darwin. (From Gager.)

ments of the 19th century; and the credit for it is due, more than to anyone else, to the great English naturalist, Charles Darwin, whose epoch-making work, the Origin of Species, published in 1859, first convinced the scientific world in general of the truth of the evolution theory.

One of the strongest indications that species are genetically connected is furnished by the resemblance in structure which is found among the animals of any group. The animals of any division of the animal kingdom are built upon the same general plan of structure, however diverse may be the modifications which they present. In the mammals the limbs, for instance, are formed after much the same pattern. In some cases bones may be fused together that are separate in other animals, or certain bones may be missing; but nevertheless it is possible by a comparative study of limb structure to show how the diverse forms may be derived from a common type. Organs which are formed according to the same fundamental pattern are called *homologous*, however diverse their form and function. On the other hand organs which perform the same function but which are different in their fundamental plan of structure are called *analogous*. Examples of the first class are afforded by the arms of man, the fore legs of a horse, the wings of bats and birds, and the flippers of the whale. These organs have been modified in various ways to subserve very different functions, but a study of their structure shows them to bear a very close resemblance nevertheless. Examples of analogous organs are afforded by the wings of birds and the wings of insects which, while they are both used as organs of flight, have very little resemblance in structure. The resemblance of fundamental plan amid differences in the way in which the plan is worked out is the natural result of inheritance from some common ancestral form, the diversity being due to adaptations to varied conditions during the divergence of species from their common ancestor.

Very striking indications of the descent of animals is furnished by the existence of rudimentary organs. These are organs of small size and degenerate structure, and they are for the most part apparently functionless. In our own bodies, for instance, there are several small muscles attached to the ear some of which are entirely functionless and others practically so. These muscles we find much more highly developed in the lower mammals where

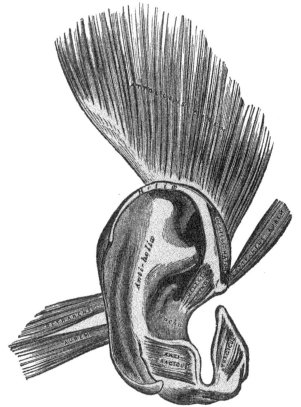

Fig. 233.—Rudimentary or vestigial and useless muscles of the human ear. (From Romanes, after Gray.)

they are of service in moving the ears. On the outer surface of the ear there is in many people a small point which represents the tip of the ear of lower mammals. The vermiform appendix which is the cause of such frequent trouble is the rudiment of an organ found in a much more highly developed state in the lower mammals where it performs a useful digestive function. The fauna of caves frequently includes many animals which are totally blind. Many of these forms have eyes in a rudimentary condition. Among the blind fishes of which there are numerous species in the caves of North America there are various degrees of degeneracy to be met with in the structure of the eyes, from those in which the eyes are fairly well developed, though functionless, to those in which they have almost entirely disappeared. The blind crayfish of the Mammoth Cave has lost its eyes, but it still preserves the eyestalks.

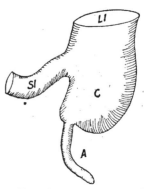

FIG. 234.—Cecum and appendix of man. *A*, appendix; *C*, cecum; *LI*, large intestine; *SI*, small intestine.

In the inner corner of our eye there is a small semilunar fold, a rudiment of the third

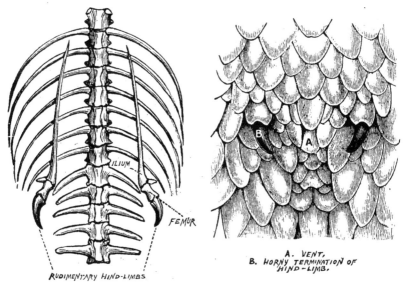

FIG. 235.—Rudimentary hind limbs of Python. (After Romanes.)

eyelid, which in many reptiles and birds is capable of being drawn over the surface of the eye. Some of the reptiles have a rudiment of a third eye which is located as a rule near the middle of the roof of the skull. In one species of reptile, Sphenodon, this eye is quite well developed, but in all other forms it exists in a very degenerate condition. This third eye was formerly connected with a part of the brain known as the pineal gland, a structure which is present, although rudimentary, in nearly all vertebrates including man. Weidersheim has recorded as many as 180 organs which are rudimentary in the human body.

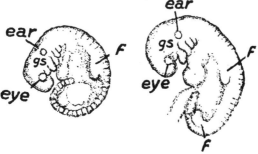

FIG. 236.—Very young human embryos showing gill slits, *gs* and rudiments of limbs, *f*. (After His.)

It frequently happens that organs which have entirely disappeared in the adult are represented by rudiments in embryonic development. The upper incisor teeth are absent in cattle, but rudimentary teeth are nevertheless found in foetal calves. In the whalebone whales teeth are no longer present, but the embryos of these whales have numerous teeth in a rudimentary condition which later disappear. All of these rudimentary organs are very naturally explained as structures which were useful once, but which have become dwindled through disuse as animals have adopted new habits of life.

The evidence of the descent of higher from lower forms furnished by embryology is often very striking. The higher vertebrates without exception show the gill clefts corresponding to the spaces between the gills of their fish-like ancestors. The arterial blood vessels of the gill region have almost precisely the arrangement found in the fishes, there being a number of aortic arches corresponding to the clefts. The arrangement is such as to carry blood to a series of gills, although no actual gills are

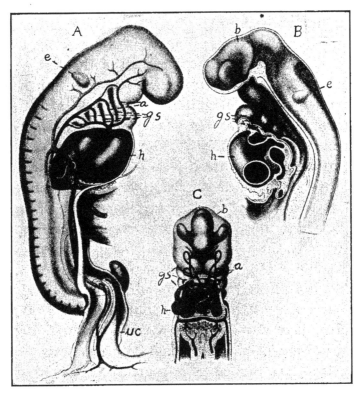

Fig. 237.—Human embryos. *A*, right side; *B*, median section; *C*, front view; *a*, arches of the aorta; *b*, brain; *e*, ear vesicle; *gs*, gill slits; *h*, heart; *uc*, umbilical cord. (After His.)

found. Later most of the aortic arches disappear, a few being more or less completely retained in the adult blood system. One of the gill clefts is modified to form the Eustachian tube which extends from the pharynx to the middle ear. The hyoid which supports several muscles of the tongue is the homologue of part of the bony framework of gill arches found in the fishes.

The student of the development of life naturally looks with interest upon the revelations of geology as to the succession of organisms on the surface of the earth. The earth itself has undergone an evolution from a relatively homogeneous condition in which it was a heated mass of molten rock. With the cooling and thickening of the crust there came the condensation of water which formed shallow seas covering most if not all of the surface. As the earth cooled further its surface was thrown into folds which gave rise to mountains. With the elevation of land there began the process of erosion and the deposition of sediment in the bottom of the primitive seas. Thus were produced the stratified rocks which have been slowly formed to a thickness of many miles. Subsequently many of the strata deposited at the bottom of the sea were raised up, thus affording to the geologist an opportunity to study the fossils or remains of living forms which they contain. It has taken a long time to trace out the succession of strata that are found, some here and some there, over the earth's surface, but the labors of many geologists have now given us a fairly adequate account of the history of the earth and its inhabitants. These rocky strata are the leaves of a great book in which the earth has written its own history. Beginning with the lowest and the earliest strata in which remains of living forms occur we can follow the successive stages in the evolution of life as we pass to more recent times. Most of the earliest records of life have been obliterated, but in the Cambrian period living forms are preserved in great abundance. Most of the phyla of invertebrate animals were represented, and there were several groups, such as the trilobites and graptolites, which have long become extinct. In the following period, the Silurian, we meet with the remains of fishes which, however, were very different from the fishes of the present day, although many of them were related to the primitive cartilaginous fishes such as the sharks and rays. During the Carboniferous era there was a great luxuriance of vegetation,

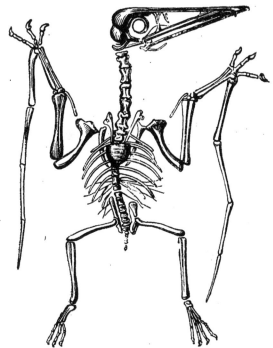

FIG. 238.—A pterodactyl, an extinct flying reptile. (After H. v. Mayer.)

mostly belonging to primitive types related to our ferns and club mosses. Much of our coal is derived from the plants of the old carboniferous forests, and we can often trace in pieces of coal the woody structure or the delicate outline of leaves of the trees that flourished at this time. In this period we find the remains of insects and amphibians, and in the following period, the Permian, we meet with those of reptiles.

It was not until the next period that the latter animals

reached the colossal size and often grotesque form that gave to this epoch the name of the age of reptiles. Great monsters stalked in the land, enormous creatures swam in the seas, and numerous queer looking pterodactyls flew in the air; but in the following epochs these great creatures all became extinct. The earliest bird of which there is record is a curious creature nearly as much reptile as bird, called the Archæopteryx. Unlike all existing birds it had a long tail with many vertebræ, and its jaws

Fig. 239.—Skeleton of a cretaceous dinosaur, *Triceratops prorsus* in the U. S. National Museum. (After Gilmore.)

were set with numerous conical teeth. It was covered with feathers, and had undoubted wings, but the wing bones were much more like those of typical fore leg than they are in our modern birds. It had been held before that birds sprang from reptilian ancestors, and the discovery of the Archæopteryx afforded a connecting link which confirmed this opinion. A few other birds with teeth are found in later strata, but their general structure

Fig. 240.—The Archæopteryx. Note the clawed digits 1, 2, and 3 of the wings, the long tail with many vertebræ, and the teeth in the jaws. (After Zittel.)

approaches more nearly that of the birds of the present time.

During the age when the reptiles were the dominant animals on the earth we find the first remains of the highest group of vertebrates, the mammals. These were represented for a long time by relatively small and primitive

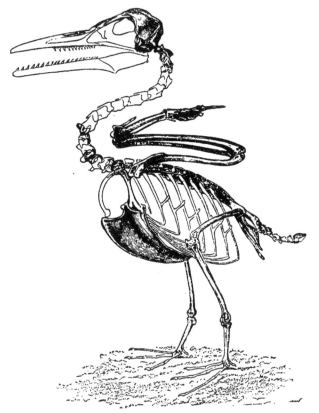

Fig. 241.—The Ichthyornis, a toothed bird of the Cretaceous period. (After Marsh.)

forms allied to the present-day marsupials. It was not until more recent times in the Tertiary period, often called the age of mammals, that the mammals became abundant. They replaced the large reptiles of the previous period, and their remains are in some cases sufficiently abundant

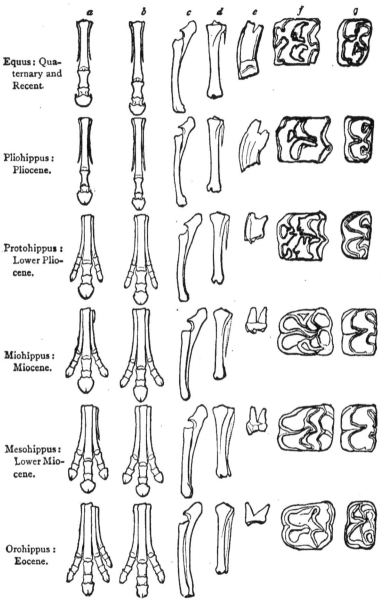

Fig. 242.—Feet and teeth in evolution of the horse; *a*, bones of forefoot; *b*, bones of hind foot; *c*, radius and ulna; *d*, tibia and fibula; *e*, roots of a tooth; *f* and *g*, crowns of upper and lower polar teeth. (From Romanes after Marsh.)

to enable us to trace the gradual evolution of some of our modern types. The history of the evolution of the horse is especially complete. In our modern horses the digits of the feet are reduced to a single one corresponding to one middle digit, the hoof representing a greatly enlarged and thickened nail. On either side of this middle digit are two rudiments, the splint bones, representing the basal part of the second and fourth digits. The first representative of the series of horse-like animals was a small creature, the Eohippus, about the size of a fox, that lived during the earliest division of the tertiary period. Its fore foot contained four toes with hoofs and a rudimentary fifth, and its hind foot had three hoofed toes. This animal was succeeded in more recent deposits by a type with four toes on the fore foot and three on the hind foot. Later appeared somewhat larger horses about the size of a sheep, with three toes on both fore and hind feet, and a rudiment of a fourth toe on the fore foot; while still later forms had but three toes on fore and hind feet, the lateral ones being much reduced in size, but still bearing hoofs. Later these lateral hoofs and their digits disappeared with the exception of the rudimentary splint bones. As the horses increased in size, the middle digit became larger and came to bear more and more of the weight of the body, while the lateral ones became smaller, and finally all but disappeared. From thirty odd species of fossil horses that are known, we can select a series of connecting links which afford the strongest evidence of the descent of our modern horses from a five-toed ancestral species.

Only in relatively late deposits do we meet with any fossil remains of man, and but very few of the oldest remains have been preserved. The oldest of all represented by the top part of the skull, a femur and a few minor fragments belong to a creature called *Pithecanthropus erectus* which was, so far as its fragmentary skeleton indicates, a human being decidedly nearer the apes than are any existing races of men. Much later we find remains of a more human type along with skeletons of the extinct cave bear, cave lion and mammoth. We find also the stone implements of primitive man, such as arrow heads and axes; at first these were crude but later they were more accurately made. These were made by men of the stone age, but later there are found implements of bronze and still later those of iron. Although man is of recent origin, geologically speaking, he has been on the earth several hundred thousand years, although we cannot measure this time with a great degree of exactness.

Taken as a whole, and despite the gaps and imperfections of the record, the history of fossil forms shows us a gradual advance from lower to higher types of life. In some cases where the record is unusually complete, as in the series of fossil horses and elephants, it enables us to follow, step by step, the evolution of our modern forms. The science of geology reveals to us an almost immeasurable past during which the seas, the continents, the mountains and the valleys of our earth were gradually being formed, and the earth's wealth of plant and animal life was gradually being evolved.

From a variety of sources, such as morphology, or the science of structure, embryology, geology, the geographical distribution of life and the observed facts of variation there is an overwhelming mass of evidence for the conclusion that plants and animals including man have arisen by a gradual process of evolution. It is a problem of great importance to ascertain by what method this great change has been effected in organic life. Ordinarily plants and animals give rise to progeny closely resembling their parents whose qualities they inherit. But occasionally organisms have been observed to depart considerably from the parental type, producing what are called *variations*. It is often remarked that no two individuals are exactly alike, and this applies as much to plants and animals as to human beings. Many of the differences between organisms of the same ancestry are due to food, climate and other environmental causes, and are probably not inherited; but other variations which occur less frequently are undoubtedly transmitted. Thus there was born in Massachusetts a peculiar ram having an unusually long body and short, crooked legs. Its enterprising owner conceived that it would be desirable to produce a breed of sheep like this ram, and he found that it was capable of transmitting its peculiar qualities to its offspring. There was thus produced the Ancon, or otter variety of sheep. The Merino sheep likewise originated from a sudden variation. Breeds of hornless cattle have been produced in a similar manner, and a great many of our cultivated varieties of plants and animals have also originated from a sudden variation. These sudden transmissible variations are commonly spoken of as *sports* or *mutations*.

It is owing to the occurrence of variations of a transmissible kind that breeders of plants and animals are able to effect such striking changes in their stock. The breeder selects those variations which are best fitted for his purpose and breeds from them. Other variations in the same direction are selected, until a race is finally produced which is often very different from the original one. Horses have been bred for speed, for strength and other qualities for generations, producing the slender high-spirited race horse on the one hand, and our heavy draft horses on the other. Cows have been bred for increased yield of milk with the

FIG. 243.—Varieties of domestic pigeons. (After Romanes.)

result that now our better breeds of cattle produce several times as much milk as did the cattle of a few centuries ago. The selections of the pigeon fancier have resulted in the production of such diverse types as the pouter, the fantail, the tumbler and numerous other varieties, all of which are considered to be the descendents of the original rock pigeon *Columba livia*. Our domestic dogs doubtless sprang from several varieties of wolf, but cross breeding and continued selection have resulted in the production of the greatest variety of form, size and disposition. Animals so different as the bull-dog, the greyhound, the newfoundland, the spaniel, and the terrier would undoubtedly have been considered members of very distinct species, if not genera, had they been met with in a state of nature. This process of *artificial selection*, as it is called, has resulted also in the production of many varieties of plants which are of the utmost value to man. By its means man has greatly increased the quality and yield per acre of his wheat, oats, corn and many other grains, and produced countless varieties of beautiful flowers. The genius of Luther Burbank has given us a stoneless plum, a spineless cactus, the Burbank potato and a large number of other improved varieties of fruits and vegetables, as well as ornamental plants.

The possibility of improving our races of plants and animals depends upon the occurrence of variations which are inherited. Variations occur in a state of nature as well as under domestication, and if there were any agency capable of selecting variations of a certain type, organisms would be modified in nature just as they have been under domestication through the agency of man. The existence of such a modifying agency was first pointed out by Charles Darwin and Alfred Russel Wallace who independently and at nearly the same time worked out their celebrated

FIG. 244.—Varieties of dogs. (After Romanes.)

theory of the origin of species by means of natural selection. According to this theory there is a process of selection continually going on in nature producing results more or less analogous to those produced by man by the process of artificial selection. Natural selection is the outcome of the struggle for existence which is ever waging in the organic world. Organisms tend to multiply so rapidly that if their propagation were not checked there would not be space enough on the earth to support them. According to Jordan a codfish may produce as many as 9,100,000 eggs per year. "If each egg were to develop, in ten years the sea would be solidly full of codfish." The elephant which is reckoned the slowest breeder of all animals would produce in 800 years, according to Darwin, 19,000,000 elephants from a single pair. In a few years more these would increase until the entire earth would be covered by elephants.

Now it is obvious that animals do not actually increase at this rapid rate. If the individuals of a species are not on the increase, as they generally are not, only two individuals from a single pair will on the average live to maturity. Their numbers are kept down by various checks such as limitation of food, climate, diseases and numerous enemies. It is only occasionally when organisms are introduced into a new country where for a time there is little to check their increase, that the high rate of multiplication, which we have described, is approximated. When rabbits were introduced into Australia and New Zealand they found few competitors and they multiplied so rapidly that they became a serious nuisance and much effort has since been expended to rid the country of the pest. Similarly when cattle were introduced into South America by the Spaniards they increased in numbers to such an extent that before many years the plains of that country were overrun by immense herds of these animals. The descendents of the English sparrow, introduced not many years ago into this country, now number untold millions.

As only a few of the descendents of any organism can as a rule survive, the chances are that, on the whole, the survivors owe their existence to the possession of some quality which gives them an advantage over their competitors in the struggle for existence. As organisms vary, those variations which are best adapted to their conditions of life will, on the whole, survive and propagate their kind. Thus there results in nature a process of selection working ever toward the preservation of the better endowed individuals. This process was called by Darwin *natural selection* in contradistinction to artificial selection which is practised by man. In natural selection we have a modifying agency which is ever tending to mould organisms into better adapted forms. In a herd of wolves for instance it would naturally work to produce greater fleetness of foot, keenness of scent, quickness of eyesight, strength, intelligence and other qualities which would give a wolf an advantage over its neighbors. According to Darwin's theory evolution has been brought about mainly through natural selection in a manner more or less similar to that in which man, by a process of artificial selection, in an infinitely shorter period of time, has been able to effect such striking modifications in his cultivated plants and domestic animals.

CHAPTER XXXIX

DIVERGENCE AND ADAPTATION

The continued modification of organisms by the agency of natural selection tends to adapt them to diverse kinds of environment. The struggle for existence is most severe between animals occupying the same region and living on the same kind of food. There is little competition between the grasshopper, the honey bee and the house fly, because they do not interfere much with one another's activities. In a small town there is competition between rival grocery stores, but comparatively little between the grocery man and the blacksmith. It is advantageous for organisms as it is for tradesmen to get their living in different ways; there is a certain escape from the rigors of the struggle for existence. Any organism which adopts a new mode of life or is able to subsist upon a different kind of food secures a certain advantage over its neighbors. As a result of this, natural selection is ever working toward the production of diversity; it tends to fill with a living organism all situations in nature which can support an inhabitant. In looking over the world one cannot avoid a feeling of wonder and surprise that Nature has filled so many different kinds of situations with living beings. She has adapted them to the severe cold of the arctic regions, to the blistering heat of arid deserts, to the depths of the oceans where many forms live in a region of cold and darkness under a pressure of several miles of water. She has modified them often into the most fantastic shapes and endowed them with most curious habits of life.

We shall consider in this chapter a few of the many adaptations which Nature has produced in order to equip her children for the great battle of life. Many animals are colored in such a way that they are difficult to detect in their natural environment. The fauna of the arctic regions contains a very large proportion of birds and mammals which are colored white like the snow and ice among which they live. Animals which live in deserts are very frequently colored much like the sand. A great many

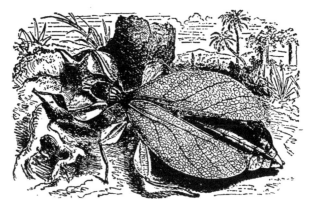

FIG. 245.—A leaf insect.

leaf-eating insects and larvæ are green, while species that are found upon the bark are more commonly of a duller hue. The celebrated leaf insect of South America has wings which simulate not only the shape and color but also the venation of leaves. One of the most striking cases of protective coloration is afforded by the Kallima butterfly which, when it is at rest on a twig with its wings folded together, closely resembles a dead leaf, the tail of its wing corresponding to the stem of the leaf, and a dark line extending across the wings presenting a close likeness to the midrib. Only the under side of the wings is colored like a dried leaf, the upper side being bright colored and conspicuous.

Contrasted with the protective coloration of many forms, are the bright colors of certain animals which are poisonous or have a disagreeable taste. Such animals are said to possess a *warning coloration.* Examples are afforded by the conspicuous spotted *Salamandra maculosa* of Europe, whose skin produces a copious supply of a virulent poison. Many poisonous snakes are very conspicuously colored and the same is true of many stinging insects. It has been found that certain non-poisonous forms closely resemble species having a warning coloration. This resemblance is called mimicry (see p. 23) and is commonly explained as having been developed because resemblance to a protected form is an advantage in deceiving enemies. A great many species of butterflies show a remarkably close resemblance to other species which are known to possess a disagreeable taste and odor.

A very common kind of adaptation is shown by the organisms called parasites. These creatures usually are carried by others from which they obtain their sustenance. We have already become acquainted with many parasitic species from various groups of the animal kingdom. In the Protozoa we have parasitic Amœbæ, parasitic infusoria and flagellates and the entire group of Sporozoa. Among the flat worms the trematodes and the cestodes, or tapeworms, are entirely parasitic. There are numerous parasitic round worms, or nematodes. There are large groups of parasitic forms among the Crustacea, such as the fish lice, whale lice and parasitic barnacles. The arachnids have their parasitic ticks and mites, and the insect world contains thousands of species parasitic on plants and animals.

Parasitism forms a relatively easy way of getting

a living and animals from a great variety of classes have taken advantage of this means of obtaining it. There is a natural check to the number of possible parasites, for if the hosts, or organisms preyed upon, were to be killed off, the parasites would starve. As it is, most animals harbor a number of these dependent creatures. Man for instance is infested with a considerable number of these parasitic Protozoa to say nothing of the numerous disease-producing bacteria. Of the trematodes, cestodes and round worms that attack him there are somewhat over fifty species. And then there are various species of ticks, mites, fleas, lice, bed bugs and other creatures which infest his person with more or less regularity according to his location or manner of life.

Parasites are classed as external, such as ticks and fleas, and internal such as tape-worms. They differ as to the degree of dependence upon their host, some, the obligatory parasites, being like the tape-worm entirely dependent on their host; others called facultative parasites being only occasionally parasitic, such as mosquitoes and biting flies. Parasitism almost always entails a certain amount of degeneration. Where parasites live in or upon their host there is often a loss of the higher sense organs, a degeneration of the nervous system, a loss of organs of locomotion or a conversion of them into organs of attachment, and sometimes a loss of the organs of digestion where the parasite lives upon the digested food of its host. An extreme case, as we have seen in the chapter on the crustaceans, is furnished by the parasite Sacculina, in which the animal has lost sense organs, appendages, digestive tract and has become converted into an irregular mass presenting no recognizable points of similarity to the barnacles to which it is related. Its life history affords an interesting illustration of the extent of degeneration to which parasitic habits may lead, as well as the importance of a knowledge of development in order to determine an animal's true affinities.

In the tape-worms degeneration has not proceeded so far, but higher sense organs are lacking and there is no trace of a digestive system. Digestive organs are quite unnecessary for the tape-worms as these animals absorb the digested intestinal contents of their hosts.

Frequently parasitic animals are compelled to live in the bodies of two kinds of host before completing their life history. The common liver fluke of the sheep, *Fasciola hepatica*, passes a part of its life history in the body of a snail before it is taken into the alimentary canal of a sheep. Most tape-worms, as we have seen, live in two different animals, usually an herbivore and a carnivore. The same is true of the trichina; and the life history of the malarial parasite is spent partly in the mosquito and partly in man. This change of host makes the perpetuation of the life of a parasite more than usually precarious. A failure to meet with either of the hosts would naturally be fatal to the parasite's career, but the increased dangers of such a method of propagation are offset by an extraordinary degree of fecundity.

Some organisms are found more or less constantly associated, although neither subsists in any way upon the other. Such forms are called *commensals, or messmates*. An example of the commensal relation is afforded by the small oyster crab, Pinnotheres, which lives between the valves of the shells of oysters and other bivalves. A third kind of association is called *symbiosis;* in this case each organism confers some benefit upon the other, so that the partnership is mutually advantageous. Many primitive organisms such as the green Hydra, the flat worm Convoluta, and many species of anemones and corals harbor multitudes of unicellular algæ, which grow and multiply within the tissues of their hosts. The algæ derive food from the carbon dioxide and probably other products of excretion from the animal's body, while they give off oxygen which is of value to the animal. Each member of the partnership thus profits by its association with the other. Another striking case of symbiosis is afforded by a species of hermit crab whose shell usually carries a specimen of sea anemone, Adamsia. The anemone secures the advantage of being carried about in situations where it can obtain more food, while by means of its nettling organs it affords a certain degree of protection to the hermit crab. If for any reason the shell of the hermit crab should be deprived of the anemone the crab hunts for another specimen which it manages to work loose from its attachment and fasten to its shell.

One of the most striking cases of mutual adaptation is afforded by the relations of insects and flowers whereby the insects are enabled to secure food and the flowers to obtain the advantages of cross fertilization. It has been abundantly shown by the investigations of Darwin in his interesting work on the effect of cross and self-fertilization of flowers that the seeds of flowers which have been cross fertilized generally produce plants of greater vigor than those resulting from self-fertilization. In higher plants fertilization is accomplished by means of pollen grains; in many plants these are blown by the wind from one flower to another, but in many others they are carried by insects. The insects attracted by the honey of the flowers become dusted with the pollen as the result of their visit and when they fly to another flower some of this pollen may be rubbed off against the surface of the stigma and thus effect cross fertilization. Many flowers show remarkable adaptations for bringing the pollen in contact with the body of the insect and also for receiving the pollen, which the insect has acquired in its previous visit to another flower of the same species. This is very well illustrated by the flowers of the sage *Salvia pratensis*. At the lower side of the opening of the flower there is a sort of platform upon which the insect alights. At the base of the flower is the nectar; the anthers or pollen-bearing organs of the flower are at the end of a movable lever which swings on a pivot. As the insect goes toward the base of the flower in the pursuit

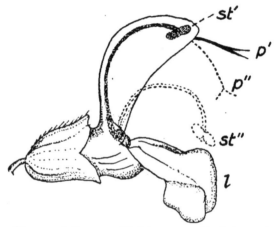

Fig. 246.—Flower of the sage (*Salvia pratensis*); *l*, lower lip of corolla; *p'*, pistil when immature; *p''*, mature pistil; *st'*, stamen when immature; *st''*, mature stamen.

of honey it causes the anther to be pushed down and some of the pollen becomes dusted upon the insect's back. The pollen and stigma mature at different periods so that a flower is prevented from becoming fertilized by its own pollen. When the stigma is ready to receive pollen it becomes extended in the position shown in the figure. When the insect visits a flower which is ready for fertilization a part of the pollen upon its back becomes brushed off upon the stigma whose position is now such as to be brought into contact with it. This is but one of the very numerous and ingenious devices for effecting cross fertilization, which we find among flowering plants. The color and scent of flowers serve to allure the insects to the places where they may secure honey. Thus both insects and plants profit by the relation. The beautiful color and sweet scent of flowers have probably been evolved in order to take advantage of the visits of insects.

CHAPTER XL

HEREDITY AND HUMAN IMPROVEMENT

Everyone is familiar with the fact that among plants and animals like begets like, or in other words offspring inherit the qualities of their parents. Everyone has noted cases of family resemblance in which certain peculiarities of form, or facial expression occur in all the children of a family. The peculiar feature known as the Hapsburg lip has characterized for many generations the members of the reigning house of Austria. A white tuft of hair has frequently been transmitted for several generations and there are numerous cases in which supernumerary fingers and toes have likewise been inherited for a long period. These are among the countless illustrations of the operation of heredity. Organisms inherit from both parents apparently to the same degree. And they also inherit qualities from their grandparents and more remote ancestors. Now and then a peculiarity which has long been latent or recessive suddenly crops out and it is then called a case of *atavism* or *reversion*.

The germ cells of organisms are the bearers of hereditary qualities. During development these cells divide and differentiate to form the various parts of the new individual, but a part of the cells of the embryo give rise to new germ cells, while others produce body or somatic cells. There is an unbroken series of cell generations from the germ cells of the parent to those of the offspring. It has been supposed by many biologists that hereditary resemblance is due to what Weismann has called the continuity of the germ plasm, or the transmission of a part of the germinal substance relatively unchanged. A part of the original germ plasm of the fertilized ovum differentiates into the bodily organs, while a part remains comparatively unchanged and forms the basis from which the new individual may arise. Parent and offspring resemble one another because both sprang from a common substance. Germ plasm comes from preceding germ plasm, the bodies of organisms being the carriers of this substance, or as Galton says the "trustees of the germ plasm." Since it has become customary to look upon inheritance as coming not from the bodies of parents but from germ cells of which the body is both the product and the carrier, many biologists think that the characteristics acquired by the body are not transmissible to the next generation. It was formerly held that those characters which an organism acquires by its own efforts or through the action of the environment were in a measure passed on to its offspring. It was thought that if a blacksmith strengthened his arm by wielding the hammer his son would have a stronger arm as a result of his father's exercise. But the majority of biologists now doubt if such acquired characters as the increased or decreased development of a particular part are ever transmitted. Numerous mutilations such as cutting off the tails of mice have been practised for many generations without producing the slightest effect upon the offspring, and certainly the decorative mutilations which many savages have indulged in for untold generations such as gashing the cheek, flattening the skull or deforming the lips, noses, ears or feet have not had the least influence on the children of these peoples who are born as free from blemishes as those of civilized man. The transmission of acquired characters is a subject on which numerous experiments have been performed, many of which have been interpreted in different ways, and there is still a difference of opinion on the question among students of heredity.

Most of the change that has been produced in domestic plants and animals has been effected, not by the transmission of characters acquired by the parents, but by means of the preservation of variations which originated in the germ cells. The appearance of six-toed cats, Ancon sheep, albinos and runnerless strawberries is due to some change in the germ plasm, and there is no doubt that such variations of germinal origin tend strongly to be transmitted. The selection of germinal variations might in time effect very great changes, and it is held by many biologists that the whole process of evolution has been brought about by this method.

It was formerly a widespread belief that various peculiarities can be impressed upon unborn children by the experiences of the mother during pregnancy. Sometimes children are born with a mark or blemish of some sort which is often attributed to a fright, desire or other strong feeling on the part of the mother. There are skin markings called nævi due to an enlargement of the cutaneous capillaries that sometimes have a certain resemblance, more often fancied than real, to strawberries, blackberries, liver or some other object for which the expectant mother may have had a strong craving. When a child is born having any sort of blemish the history of the mother is inquired into to discover some experiences which might account for it, and out of the numerous experiences that have occurred something is frequently found that satisfies the enquirer. Stories like the following are typical: Mrs. A. on putting her hand into a flour bin was frightened by a mouse which ran upon her arm. Her child born some weeks afterward had a reddish patch on its arm, resembling a mouse. In another case a dentist raised up the lip of a lady to look at an eye tooth. Her child born shortly afterward had a hare lip or lip cleft at one side like that of a hare. The first case was probably a coincidence and the second cannot have been due to the alleged cause because hare lip, a phenomenon occurring in lower animals as well as in man, is due to lack of junction of two embryonic rudiments and is caused very much earlier in development than the period in question. The connection between mother and child is established through the organs of circulation and while anything such as sickness, starvation or alcohol which deteriorates or poisons the blood of the mother may be very unfavorable to the child there is no good evidence that the mother's imagination can paint pictures on the child's body, or bring about specific deformities. Most of the latter which are attributed to maternal impressions, occur also in the lower animals and are well known to pathologists as due to quite different causes.

Heredity is fast coming to be the subject matter of an exact science. This is largely due to the discovery of a remarkable law which is named after its discoverer, Gregor Mendel, an Austrian monk. For years Mendel had been experimenting by crossing varieties of sweet peas and other flowers; he finally published his results in a rather obscure periodical where his papers remained unnoticed until the year 1900 when they were brought to light. The rediscovery of Mendel's law in 1900 made an epoch in the study of heredity, for those who have followed in Mendel's footsteps, verifying and extending the investigations which he began, have shown that Mendel's long forgotten law affords the key that unlocks many mysteries previously obscure and makes it possible to attain results in experimental breeding of the greatest practical value.

Let us see what this law is. In crossing varieties of peas that differed in certain well-defined characters Mendel found that in the first generation the offspring were not intermediate in respect to the characters in question, but that one character was represented apparently to the complete exclusion of the other. Thus crosses of tall and dwarf peas produced nothing but tall peas, and crosses of

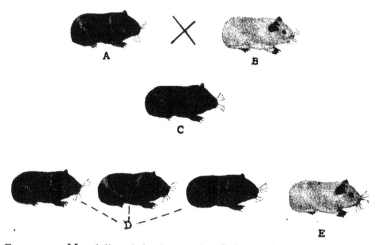

Fig. 247.—Mendelian inheritance in Guinea pigs. A black animal, *A*, mated with an albino, *B*, produces progeny which are all black like *C*. These progeny have albinism in a recessive state, and when they are mated produce blacks, *D*, and albinos, *E*, in the ratio of three to one.

yellow with green peas produced only peas that were yellow. The character such as tallness or yellowness which appeared to the exclusion of its opposite was called *dominant*, the suppressed character being called *recessive*. The most striking results, however, were obtained in the second generation after the members of the first generation were crossed with one another or self-fertilized. The recessive character was then found to reappear in one-fourth of the progeny, the other three-fourths showing the dominant character. The dwarf peas thus arising produced nothing but dwarfs, the dominant character never reappearing. It was found also that one-third of the tall peas produced nothing but tall plants, the recessive character never appearing. These were called the pure dominants; two-thirds of the tall peas, however, when interbred produced tall and dwarf in the ratio of 3 to 1. Two-thirds of the talls were therefore mixed as regards the tall and dwarf characters. It is customary to write a formula for the second generation of hybrid crosses in the following way: $1DD + 2DR + 1RR$. In the case of our second generation of peas one-fourth would be pure dominants or DD's, two-fourths impure, DR's, and one-fourth pure recessives, RR's. Similarly the progeny of the yellow peas resulting from crosses of the yellow and green varieties are found to produce yellow and green peas in the ratio of 3 to 1. Of the yellows one-third were pure yellows, and two-thirds produced both yellow and green peas, whereas the greens being the pure recessives produced nothing but green peas. The most significant thing about Mendel's law is that contrasted characters come to be segregated out in the second generation of hybrids in definite numerical ratios. Characters such as greenness, yellowness, tallness, shortness and many others apparently behave as units capable of being combined and separated again without losing their identity.

Where organisms differing in two pairs of contrasted characters are crossed each pair behaves as a rule independently of the other. Thus when a tall, yellow pea is crossed with a dwarf green variety the first generation is tall and yellow; the second generation is constituted as follows: 9 tall yellow : 3 tall green : 3 dwarf yellow : 1 dwarf green. Where several pairs of characters enter into the combination the ratios are still more complex.

Mendel's law has been found to apply to a great many plants and animals. The breeder who knows the ancestry of his stock is able to anticipate what will be the various forms that will probably arise from certain matings. He is often able to effect new combinations of qualities and the practical value of such knowledge applied to the improvement of grains, fruits, vegetables and domestic animals has already amounted to millions of dollars for farmers and stock raisers.

But one of the most important applications of Mendel's law is in the field of human heredity. Heritable human traits have been studied very extensively in recent years with the result of throwing a flood of light on many subjects hitherto obscure. Eye color in man as in animals is one of the many characters that has been found, in most cases at least, to obey Mendel's law. Dark colors are dominant over lighter ones. Matings of two black-eyed people may produce only black-eyed children or again some of the children may have brown, gray or blue eyes. The lighter colored eyes are due to the fact that the parents were not pure dominants, DD's, but mixed, DR's. Matings of dark- and light-eyed people may produce both dark- and light-eyed children, but matings of two light-eyed people cannot be expected to produce a dark-eyed child, as the light colors are recessive to dark. When both parents have blue eyes all of the children will almost always have blue eyes also. There is evidence of Mendelian segregation in hair color, although the latter is subject to much variation through environmental causes. Dark colors are dominant to light and consequently while matings of dark-haired parents may produce some light-haired children, light-haired parents would not be expected to produce dark-haired children. Albinism, a condition in which hair and eyes are devoid of pigment, is recessive in man as in animals. Two albinos will produce nothing but albino offspring.

Deafness is sometimes a hereditary and sometimes an acquired character. When it is due to accident or disease there is little danger of its transmission, but when it is caused by some inborn defect it is very apt to afflict the following generation, especially if both parents are congenitally deaf.

It often happens that there may be a strong proclivity toward some defect or disease which may be overcome

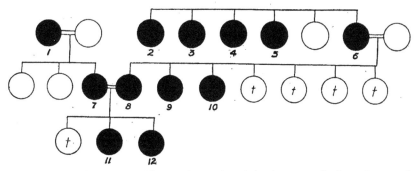

Fig. 248.—Diagram illustrating the inheritance of insanity and other forms of mental defect. 1, alcoholic; 2, feeble minded; 3, feeble minded; 4, hypochondriac; 5, had insane daughter; 6, visionary, drunken wreck; 7, eccentric; 8, insane; 9, unbalanced; 10, crazy, fits of temper, gets wild; 11, insane; 12, microcephalic, defective, died in infancy; † died in childhood. (From data from Rosanoff and Orr.)

by wholesome living under favorable conditions. It is commonly held that people may inherit a tendency toward tuberculosis, gout, Bright's disease and numerous other maladies; but if a person takes the proper precautions he may avoid these impending dangers. Knowledge of hereditary predispositions is often of great service to the physician in understanding his patient's case, especially if he is dealing with nervous or mental ailments. That heredity plays an important rôle in the production of insanity has long been recognized, but it is only recently that evidence has been brought forward to show that insanity is inherited in accordance with Mendel's law. We should distinguish hereditary insanity from insanity that may be the result of accident, disease, alcoholism or severe shock, although in many cases these alleged causes only serve to awaken a hereditary proclivity that is not strong enough of itself to make the person insane. Insanity is not inherited in so clear cut a manner as albinism or eye color; what is transmitted is rather a nervous condition that may manifest itself as insanity, epilepsy or other nervous disorders. Insanity appears to behave as a recessive or partially recessive character. --Matings of normal and insane frequently produce only normal children, although if the normal partner came from insane ancestry a part of the children would probably be insane. When both parents are afflicted with hereditary insanity all of the children may be expected to become insane or have some severe nervous affliction. If both parents are sane but of insane ancestry the expectation is that one-fourth of the children will be afflicted. While there may be no cases of insanity among the immediate relatives of the afflicted person his insanity may nevertheless be inherited from more remote ancestors. From what is known of the inheritance of insanity it is clear that people in whom insanity is an inherited trait should not marry and incur the risk of transmitting this terrible malady to their children.

Several careful studies have shown that feeblemindedness, like insanity, is inherited as a Mendelian character. Feeblemindedness is capable of being detected in early years, before insanity usually manifests itself, and in the majority of cases it is clear that it is an inherited character, although it sometimes results from sickness or injury. A very interesting study of the inheritance of feeblemindedness is contained in a little book by Goddard,

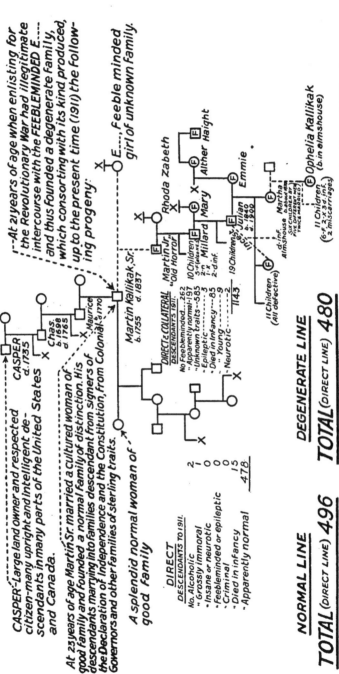

Fig. 249.—Chart of the Kallikak family. (From a diagram issued by the Eugenics Record Office.)

entitled "The Kallikak Family," in which there is traced the pedigree of several hundred descendants of one feebleminded girl. Martin Kallikak, a revolutionary soldier, had by this girl an illegitimate and feebleminded son, who raised a large family most of whom were feebleminded. Altogether there were traced the histories of 480 direct descendants of the feebleminded girl. Of these 143 were definitely known to be feebleminded while the mental condition of 291 was doubtful. Thirty-six of the children were illegitimate and 82 died in infancy; 33 descendants were sexually immoral, 24 were confirmed drunkards, 8 keepers of disreputable houses; 3 were criminals and 3 epileptics. Only 46 were known to pass as normal. In the Nam family studied by Davenport about 90 per cent. were feebleminded, and about 90 per cent. of the men were addicted to alcohol. One-fourth of the children were illegitimate and the history of the family is one of pauperism, debauchery and crime. The cost of this family to the community is estimated at a million and a half dollars.

While we cannot properly speak of crime itself as inherited, criminal tendencies are known to run through families, some being addicted to thieving, others to crimes of violence, etc. In the notorious Jukes family whose members to the number of 2094 have been traced through several generations there were 181 victims of intemperance, 299 paupers, 118 criminals, 378 prostitutes and 86 keepers of brothels. In one branch of the family taking its origin from a disreputable woman known as Margaret, the Mother of Criminals, there have been over 800 descendants including numerous paupers, criminals and prostitutes.

A very large proportion of the criminal class, but by no means all, are of defective mentality. Fernald states that in Massachusetts "at least 25 per cent. of the inmates of our penal institutions are mentally defective." Criminals show a high percentage of epileptics who are especially predisposed to crimes of violence. Reformatories and homes for delinquent girls show that a remarkably large number of these offenders are of defective intelligence, the proportion in an institution at Geneva, Illinois being 89 per cent. The Massachusetts Commission for the investigation of the White Slave Traffic reports that of 300 prostitutes, 154 or 51 per cent. were feebleminded, and most of the others "were of distinctly inferior intelligence." Besides those who are feebleminded there is a large element of the population who, born with inferior intellect and weak will, cannot secure or retain positions of responsibility, and who are unable to resist the temptations of the evil surroundings amid which they are prone to drift, and who are apt to find their way sooner or later into the ranks of the tramps, paupers, chronic alcoholics or criminals.

Our defective classes are a great drag upon society and a menace to its peace. They necessitate homes for the feebleminded, reformatories, homes for delinquent girls, asylums for the epileptics and insane and they help to fill our jails and penitentiaries. But the millions spent upon those maintained in institutions is a small part of their total cost to the country. The larger part of those who are defective mentally are unconfined and they help to swell the number of incompetents, vagrants, criminals and immoral women whose evil influence through the spread of loathsome diseases is beyond calculation. Could society rid itself of its hereditary defectives there would result directly and indirectly untold benefits; there is no social question more urgent than how the race can purge itself of its undesirable stock.

In the production of defective human beings we must not forget the direct influence of the environment. Where children are raised under conditions of extreme poverty, crowded together in unsanitary tenements with a scant allowance of poor food, and breathing impure air, the deteriorating effects upon both mind and body generally handicap the individual for life. Undoubtedly the conditions under which Kallikaks, the Jukes and other notorious families have lived contributed much to the deterioration of their members, but it would be a great error to attribute their degradation to environment alone. What an individual becomes is a product of his inheritance on the one hand and his environment on the other. Where either is very bad his chances of amounting to much are greatly reduced; but it better to have a bad environment than bad inheritance, for many individuals born with good stuff in them have risen to eminence, despite all sorts of obstacles that surrounded them. Sanitary surroundings, education and a wholesome moral atmosphere might have made passable citizens out of many of the Kallikaks and Jukes, but these advantages probably would not have made any of them a great discoverer or President of the United States. It is with human beings as it is with varieties of corn. Some kinds of corn will give rise in ordinary soil to large fine ears while others raised under identically the same conditions will produce but a scanty yield of small ears which are poorly filled. The poor scrubby variety with superior cultivation may be made to give an increased yield, but no farmer would think of planting it if he could get seed of the better kind.

Among men as among lower animals and plants it is the breed that tells. But among men bad inheritance tends to make bad environment, for the naturally weak and incompetent and the vicious tend to associate with their like and to form a social stratum where vice flourishes and crime is bred. Does bad environment in turn produce bad inheritance? This is a very important question upon which it is very desirable that more should be known. Some writers have answered it in the negative, but several recent experiments have shown that the question may have to be answered in the other way. In his experiments on the hereditary influence of alcohol in Guinea pigs Stockard found that the offspring of animals subjected for several weeks to the fumes of alcohol were frequently undersized and of diminished vitality. The animals tested were first bred together and found capable of producing healthy young. Afterward they were kept under the influence of alcohol, and then bred and the young compared with those of the former matings of normal Guinea pigs. Matings were made between normal males and alcoholic females, normal females and alcoholic males and alcoholic males and alcoholic females. All of these gave a considerable number of offspring that were dead at birth (still born) and several young dying soon after birth, the proportions in both cases being highest in those matings in which both parents were alcoholic. The striking difference between progeny of alcoholic and normal parents is shown in the following table:

	No. of matings	Neg. results, or early abortions	Young soon dying	Survivors
Alc. ♂* × normal ♀	59	25 matings	8 litters	33
Alc. ♀ × normal ♂	15	3	3	10
Alc. ♂ × alc. ♀	29	15	3	9
Summary	103	43	14	56
Controls	35	2	1	56
2d generation:				
Alc. ♂ × alc. ♀	19	7	c	13
Normal ♀ × alc. ♂	3	....	2	2

* ♂ = male; ♀ = female.

Several of the animals resulting from alcoholized parents were bred together, and although they had themselves not received any alcohol, they produced a high percentage of weak and defective offspring. Apparently alcohol produced an effect on the germ cells that caused a hereditary weakness in the offspring that was handed on after the original cause was removed. The parent animals showed little effects of alcohol. There was no transmission of an acquired character, but the production of a germinal variation by the action of alcohol on the germ plasm.

These results make it probable that in human beings also alcohol may be a potent cause of defective inheritance, as it has long been regarded by many physicians and physiologists. Idiocy, feeblemindedness and epilepsy are found with exceptional frequency among the children of people addicted to alcohol. On the other hand, chronic alcoholism is in a large proportion of cases the result of inherited defects, so that the children of alcoholics may show weakness because their parents were defective quite aside from their alcoholism. What starts defective strains of humanity in the first place we do not know, but it is probable that much of the mischief must be laid to the score of alcohol.

It is of the highest importance to our race that we abolish the sources of our hereditary feeblemindedness, insanity, epilepsy and criminality. It is also of the highest importance and an imperative social duty to prevent the propagation of the hereditary defectives that we have. This class, speaking generally, is unusually prolific. Although infant mortality is high among them, it is the opinion of many of the foremost students of the problem that our defectives are on the increase.

The feebleminded as a rule are remarkably prolific.

Speaking of conditions in England Whetham remarks: "The workhouse records frequently note that five, six, or seven children have been born before the mother is twenty-five years of age, and she herself may have commenced child-bearing at fifteen years of age or even younger. Most of these children inherit the mental condition of their parents, and where both parents are known to be feebleminded, there is no record of their having given birth to a normal child. In one workhouse there were sixteen feebleminded women who had produced between them 116 children with a large proportion of mental defect. Out of one such family of fourteen, only four could be trained to do remunerative work.

"With regard to the fertility of feebleminded stocks, it has been pointed out that the feebleminded children from the degenerate families who use the special schools in London, come, sometimes two or more at a time, from households averaging about seven offspring, whereas the average number of children in the families who now use the public elementary school is about four."

Most of the feebleminded are not confined in institutions. Of the 10,000 known cases in Pennsylvania, according to Dr. Barr, 6500 are free. In many cases women in almshouses continue to produce children most of whom are below par mentally. In England a few years ago the girls born in workhouses were set adrift at sixteen. They frequently returned repeatedly to give birth to illegitimate children who were raised at the county's expense until old enough to follow the mother's example, an effective system for encouraging the increase of undesirable stock.

The race can improve its heritable qualities only by breeding from the best and preventing the breeding of its worst. To bring this about so far as it is possible is the practical aim of the science of Eugenics. At the present time there are various agencies working toward the deterioration of the inborn qualities of the race and others which have the opposite tendency and the trend of our biological development depends upon which set of tendencies is the more potent. One very serious condition which has come into existence largely during the last half century is the diminishing size of the family among the educated and successful classes. While there has been a general decline in the birth rate in most civilized countries this decline does not affect all classes of society alike. A decreasing birth rate is in itself no menace so long as it is not so low as to cause, as in France, an actual decrease of population. But when the defective classes and those with inferior inheritance continue to produce many children while the birth rate of those with superior endowments becomes greatly reduced, a deterioration of the hereditary qualities of the race will inevitably result.

Seventy-five years ago the educated classes had as a rule large families, but their birth rate has gradually decreased until they average in the United States between two and three children. The average number of children of graduates of Yale and Harvard is scarcely over two per family and the families of most other college graduates are but little larger. It is much the same with successful men of business and the higher classes of artizans. The birth rate of foreign immigrants is much higher, the average family in Massachusetts for instance having 4.7 children while the families of the American born have only 2.7 children. It requires nearly four children per family to replenish the population; many fail to marry or else die before reaching maturity so that, starting with four, only two (or a little over) take the places of their parents and become producers of children. It follows that stocks with smaller families of two or three children are bound to become extinct in a few generations. At the present time we are rapidly losing the blood that formerly gave us our scholars, legislators and leaders of men. While people of very moderate natural abilities are found in the so-called educated classes, there is little doubt that the latter have an average inheritance above that of the general population. But however this may be, it is the duty of those who have been blest with good hereditary qualities to see that their race does not perish. If such people shirk the responsibilities of parenthood, as they are nowadays so prone to do, and leave the perpetuation of the race to inferior strains who tend to be only too prolific, the race cannot fail to deteriorate. Man, to a great extent, has abolished the struggle for existence, and now his future lies largely in his own hands to make or to mar.

INDEX

A

abdomen, of grasshopper, 2, 5
    of crustaceans, 84–88, 91
Acridiidæ, 18, 19
adductors, 98
adrenalin, 281
adrenals, 281
air bladder, 160, 167, 168
air cells, 273
air sacs, 188
albinos, 372, 376, 377
alcohol, how made, 237, 238
    effect on digestion, 257, 258
      on heart and circulation, 268
      on kidneys, 279
    general effects, 311–314
    hereditary effects of, 383–384
alimentary canal, of bee, 58
    of bird, 187
    of crayfish, 86, 87
    of grasshopper, 6, 7
    of man, 249–251
alligators, 176, 180
alternation of generations, 135
Amazon ant, 68
Amia, 161, 167, 338
ammonites, 105
Amœba, 146, 147, 151, 240, 241, 331
Amphibians, 157, 169–175
Amphioxus, 155, 156, 337
amphipods, 94
amylopsin, 251
anaconda, 179
anal fin, 158

analogous organs, 343
Ancon sheep, 356, 372
anemones, 130, 136
Anguilla, 163
Annelida, 114
Anopheles, 47
ant eater, 214, 215
Ants, 36, 56, 65–69
antelopes, 222
antenna cleaner, 58
antennæ, 3, 68
anthropoids, 227
antitoxins, 326
antlers, 221, 222
Anura, 169–175
anvil, 310
aorta, 263
    arches of, 347
apes, 227, 230
aphids, 35, 36, 68
aqueous humor, 307
Arachnida, 18, 76–83
Archæopteryx, 350, 351
Argentine ant, 69
Armadillo, 214, 215
arteries, 88, 100, 262–265, 267
artificial selection, 356–359
Artiodactyla, 220
Ascaris lumbricoides, 122
Ascaris megatocephala, 122
Asellus, 93
assassin bugs, 32
assimilation, 239, 252
asteroids, 108–110
astigmatism, 308
atavism, 370

atoms, 233
auricles, 100, 262
axis cylinder, 296

B

bacillus, 317, 323–326
Bacillus tuberculosis, 322, 323
backswimmers, 33, 34
bacteria, 150, 261
balancers, 40
Balanoglossus, 155
barnacles, 94–97
Basilarchia archippus, 23
bass, 167
bathing, 286
Batrachians, 157, 169–175
bats, 185, 218
beak, 84
bears, 223
beavers, 216, 217
bedbugs, 32
bees, 56–64
beetles, 18, 51–55
Behring, von, 326
bile, 251
birds, 183–208, 338–340, 350–352
birth rate, 386
bladder, 277
bladder worm, 127–129
Blattidæ, 18, 19
blood, of clam, 100
  of crayfish, 88
  of earthworm, 117
  of grasshopper, 7, 8
  of man, 259–265
blow flies, 42
boa constrictor, 179
boils, 321
Bombyx mori, 26, 27
bones, 287–291
bot flies, 43
Bovidæ, 221
brain, of Amphioxus, 155, 156
  of clam, 100, 101

brain, of crayfish, 88
  of earthworm, 118
  of grasshopper, 8
  of man, 230, 231, 294, **295**, 298–303
branchiæ, 85, 86
Bright's disease, 279, 377
brittle stars, 110
bronchi, 273
Buffalo gnats, 44
Bufonidæ, 173
bugs, 31–34
bull-frog, 172, 173
bumble bees, 63
Burbank, L., 358
Burroughs, J., 207
byssus, 102

C

cabbage butterfly, 20, 21
caddis flies, 70, 73
calciferous glands, 117
camel, 220
canine or eye teeth, 247
capillaries, 264–266
carapace, 84, 86
carbohydrates, 237, 249, 251, 255
carbon, 233, 234, 235, 236
carbon dioxide, 236, 238, 239, 240, 269, 271, 273, 275, 276
Carboniferous period, 349
Carnivora, 223
Carolina locust, 12
carpals, 290
cartilage, 243, 287
casein, 260
cats, 207, 208, 223
cells, 240–244
centipedes, 75, 76
Cephalopoda, 105–107
Cephalothorax, 84, 88
cerebellum, 299–301
cerebral hemispheres, 300–303
cerebro-spinal system, 296

Cervidæ, 221
Cestodes, 126-129
Cetacea, 225, 226
chalk, 148
chambered nautilus, 105, 106
chameleons, 179
cheese mites, 82
Cheiroptera, 218
chelæ, 85
chelipeds, 85
Chelonia, 176, 181
Chelonians, 176, 181, 182
chemical changes, 232-234, 245
chimpanzees, 227, 229, 230
chinch bug, 31, 32
chitin, 2
chitons, 105
chlorine, 236
chlorophyll, 145
chocolate, 257
Chordata, 156
choroid coat, 307
cicada, 34, 35
cilia, 242
clams, 98-102, 151
class, 18
classification, 17-19
clavicle, 290
cleavage, 336
clitellum, 115, 119
clot, 259
clothes moths, 30
clotting of blood, 260.
cobra, 179
Coccidæ, 36-38
Coccidea, 148
cochlea, 310
cockroaches, 15, 18, 19
cocoons, of insects, 26, 27, 67
    of earthworms, 119
    of spiders, 77, 338
cod, 167, 339, 360
codling moth, 29
Cœcilians, 169, 180

cœcum, 250, 345
Cœlenterates, 130-139
Cænurus çerebralis, 129
coffee, 257
cold spots, 305
colds, 275, 320-323
Coleoptera, 18, 51-55
Colorado potato beetle, 52, 53, 204
comb, 58, 60, 61
commensals, 366
commissures, 8, 100, 101
compounds, 232
conjugation, 143, 332
connective tissue, 242, 243
consumption, 275
contractile vacuoles, 142, 147
convolutions, 300
copepods, 97, 151
copperhead, 179
coral polyps, 130, 136-138
coral snake, 178
corals, 136-138
corium, 283
cornea, 3, 307
corpus callosum, 300
corpuscles, of blood, 259, 261, 265, 266
cortex, 300
cottony-cushion scale, 38
coughing, 299
courtship, 80, 160, 194-197
crabs, 91-93
cranium, 160, 289
crayfish, 84-90
cremaster, 21
cretinism, 281
crickets, 14, 15, 18, 19
Crinoids, or sea lilies, 112
crocodiles, 176, 180, 181
Crocodilia, 176
crop, of birds, 187
    of earthworm, 116
    of grasshopper, 6
    of honey bee, 58

INDEX

cross fertilization, 367-369
Crustacea, 18, 84-97, 337
crystalline lens, 307
Ctenophores, 139
cuckoos, 198
curculios, 55
cuticle, 283
cutis, 283
cuttle fish, 106
Cyclostomes, 156, 157
Cynipidæ, 69
cysticercoid, 127
cysticercus, 127
cysts, 144, 147

D

damsel flies, 70, 72
Daphnia, 97
Darwin, C., on cross and self-fertilization, 367
    on earthworms, 119
    on natural selection, 358, 360
    on sexual selection, 195, 196
    on species, 17, 342
Davenport, C. B., 380
deafness, 377
Decapoda, 93
deer, 220-222
degeneration, 96, 365, 366
Demodex, 83
dentine, 248
devil-fish, 98, 107
dextrose, 251, 254, 266
diaphragm, 274
digestion, in Amœba, 147
    in birds, 187, 188
    in crayfish, 87
    in grasshopper, 7
    in Hydra, 131
    in man, 245-251
    in Paramœcium, 142
    in starfish, 109, 110
diphtheria, 326

Dipnoi, or lung fishes, 160, **167**, 168
Diptera, 18, 40-50
disinfectants, 320
Dissosteira, 18
dobson, or hellgrammite, 74
dogfish, 161, 166, 167
dogs, 210, 223, 358, 359
dorsal fins, 158
dorsal vessel, 117
dragon flies, 70-72
drones, 56, 60
drum membrane, 309
duck bills, 211, 212
ductless glands, 280-282
Dytiscidæ, 55

E

ear, 309, 344
earthworms, 114-120, 276, 334
Echidna, 212
Echinococcus, 129
Echinoderms, 108-113
Echinoids, 110-113
ectoderm, 131, 139, 336
ectoplasm, 147
Edentata, 214, 215
eels, 163, 167
eggs, 233-236
    care of, 337-340
    of amphibians, 170, 171
    of ants, 67
    of birds, 188, 189, 191, 192, 198, 206
    of bot-fly, 43
    of butterfly, 21
    of clam, 101
    of cockroach, 15
    of crocodiles, 181
    of crustaceans, 85, 90, 96
    of dragon flies, 70
    of earthworm, 118, 119
    of fishes, 160, 238, 239, 161, 163, 164, 166

eggs, of fleas, 49
   of fluke, 125, 126
   of grasshopper, 10
   of hookworm, 123
   of house-fly, 41, 42
   of Hydra, 130, 132, 133
   of jelly fish, 135
   of katydid, 14
   of mosquitoes, 45, 46
   of moths, 28–30
   of spiders, 78, 80
   of sponges, 140
   of tapeworm, 127
Elasmobranchs, 164–166
electric-light bugs, 33
elements, 232
elephants, 220, 360
elk, 222
embryology, 334, 346, 347
enamel, 248
endoplasm, 142, 147
English sparrow, 205, 206
entoderm, 131, 133, 139, 336
enzymes, 238
Ephemeridæ, 72
epidermis, 283
epiglottis, 272
epilepsy, 303, 378, 384
epinephrin, 281
epiphragm, 104
epithelium, 244, 283
esophagus, of birds, 187
   of crayfish, 87, 88
   of earthworm, 116, 117
   of grasshopper, 6
   of man, 248, 249
Eudorina, 333
Eugenics, 386
Euglena viridis, 144, 145
Eustachian tube, 310, 347
evolution, 341–369
excretion, 8, 239, 277–279
exercise, 292, 293
expiration, 274

eyes, of Amphioxus, 156
   of crustaceans, 84, 88, 89, 93, 94
   of grasshopper, 2, 3
   of man, 305–309
   of snail, 103
   of spiders, 76
   of squid, 106
   rudiments of, 349

F

family, 18
Fasciola hepatica, 124–126
fats, 237, 249, 251, 255
feathers, 183–185, 195–197
feeblemindedness, 378–381, 384, 385
Felidæ, 223
femur, of grasshopper, 4
   of man, 290
fermentation, 237, 238
ferments, 238
Fernald, G. G., 380
fertilization, 160, 163, 164, 334, 367–369
fiber, 243, 244, 291
fibrillæ, 243
fibrin, 259
fibrinogen, 260
fibula, 290
Fishes, 48, 157, 158–168
fission, 143, 145, 147, 331
flagella, 139, 141, 144–146
Flagellata, 141, 144–146
flat worms, 122, 124–129
fleas, 48, 49
flesh flies, 42
flukes, 124–126
foot, 99, 103, 105
Foraminifera, 148
Forbush, E. H., 204
frogs, 169, 171–175
fungi, 317

G

gall, 251
gall bladder, 251
gall flies, 69
galls, 69
Galton, F., 371
ganglia, 8, 9, 88, 99-101, 118, 296
ganglion cells, 296
ganoids, 166
gastric cœca, 6
Gastropoda, 102-105
gastrula, 336
gavial, 181
Geiger, J. C., 327
genus, 17, 18
geology, 341, 348, 355
germ cells, 370
germ plasm, continuity of, 370, 371
germicides, 320
gibbons, 227
Gila monster, 180
gill-slits, 154, 155, 159, 346, 347
gills, 276
    of amphibians, 170
    of Amphioxus, 155
    of crayfish, 85, 86
    of cyclostomes, 157
    of fishes, 159
    of May-fly larvæ, 72
    of mollusk, 98, 99, 103, 104, 106
gipsy moth, 28
gizzard, 116, 187
glomerulus, 278
glycogen, 251, 254, 255
Glyptodon, 215
Goddard, H. H., 378
goiter, 281
Goltz, F., 301
Gordius, 124
gorillas, 227, 230
Grantia, 139
grasshopper, 1-14, 17-19
green gland, 84, 88
Gregarines, 148
Grillidæ, 18, 19
grizzly bear, 224

H

Hæckel, E., 114, 155
hæmoglobin, 260, 270
Hæmosporidia, 148-150
hag-fishes, 156, 157
hair, 209, 284, 285
halteres, 40
hammer, 310
harvest fly, 34
hawks, 204, 205, 338
hearing, 6, 304, 309, 310
heart, 7, 8, 88, 100, 154, 156, 262-268
hearts, of earthworms, 116, 117
heat spots, 305
hedgehogs, 209, 216
Helix, 103
Helmholtz, H., 313
Hemiptera, 31-39
heredity, 370-387
hermaphrodites, 334
hermit crabs, 40, 91, 367
Herrick, F. H., 191-194
Hessian fly, 45
Heteroptera, 31-34
hibernation, 171
hippopotamus, 219, 220
Hodge, C. F., 311
holophytic forms, 144
Holothurians, 112
Homarus americanus, 90
Hominidæ, 230
Homo, 230, 231
Homo sapiens, 231
homologous organs, 343
Homoptera, 35-39
honey-sac, 58
hookworm, 123, 124

INDEX 395

hopper-dozers, 14
horned toads, 179, 180
hornets, 65
horse, evolution of, 353, 354
    improvement of, 356
horse flies, 44
horse-hair snake, 124
house fly, 40, 41, 42
Hudson, W. H., 191
humerus, 290
Huxley, T. H., 84, 230, 238
Hydra, 130-133, 240, 241, 276
Hydra viridis, 133
hydrogen, 234, 236-238
hydroids, 130, 133-135
Hydrophilidæ, 55
hydrophobia, 326, 327
Hyla, 175
Hylidæ, 174, 175
Hymenoptera, 56-69, 338
hyoid bone, 347, 348

I

ichneumons, 69
imago, 12, 73
immunity, 326
incubation, 188-190
Infusoria, 141-144
insanity, 303, 377, 378
Insectivora, 216
inspiration, 274
instinct, 190, 191
intelligence, seat of, 300-303
internal secretion, 280-282
intestine, of clam, 100
    of crayfish, 87
    of earthworm, 116, 117
    of grasshopper, 7
    of man, 249, 251
Invertebrates, 153, 348
iodine, 236
iris, 307
iron, 236

Isopoda, 93
itch mites, 83

J

jaws, 76, 84
jelly fish, 130, 134, 135
Jenner, E., 328
Jordan, O. S., 167, 168, 360
Jukes family, 380, 382
June-bugs, 51, 53

K

Kallikak Family, The, 379, 380, 382
Kallima butterfly, 363
kangaroos, 213
katydids, 13, 14
kidneys, 8, 100, 277-281
kissing bug, 32
Kitchener, 312
Koch, R., 323
Koebele, 38
Kræpelin, 312

L

labial palpi, of grasshopper, 3, 4
    of clam, 99, 100
labium, 4, 31, 58
Lacertilia, 176
lachrymal glands, 306
lacteals, 266
lady beetles, 54
lamellæ, 98
lamellibranch, 98
lampreys, 156, 157
lancelet, 155
larvæ, of amphibians, 175
    of clam, 101
    of echinoderms, 113
    of tunicata, 154
    of insects, 12, 20, 21, 26, 28, 29, 30, 41-49, 51, 52, 62, 66, 67, 70-74
larynx, 272

Latrodectes, 80
leaf insect, 15, 363
leeches, 120, 121, 334
lemurs, 227
lens, 307, 308
Lepidoptera, 20–30
leucocytes, 260, 261, 265
ligaments, 287
Limnæa, 104
Limnoria lignorum, 93
lions, 223
lipase, 251
liver, of clam, 100
 of crayfish, 86, 87
 of man, 250, 251–254, 255, 257, 277, 279, 280
liver fluke, 124–126
lizards, 176, 179, 180
lobster, 90
Locustidæ, 18, 19
Loeb, J., 334
Loligo pealii, 106
Lubbock, Sir J., 66
Lumbricus terrestria, 114–120
lung books, 77
lung fishes, 160, 167, 168
lungs, 103, 168, 170, 270–276
lymph, 266
lymphatics, 266

M

mackerel, 167
malaria, 47, 48, 148, 150
Malpighian corpuscle, 278
Malpighian tubules, 7, 8
Mammalia, 209–231, 338, 339, 340, 352
Mammals, 209–231, 338, 339, 340, 352
mammary glands, 209, 339
mammoth, 355
man, 227, 230, 231, 355
mandibles, 3, 31, 58, 84

Mantidæ, 16, 19
mantids, 16
mantle, 98
manubrium, 134
marrow, 289
Marsupialia, 212–214
maternal impressions, 372, 373
mating, 194–197
maxillæ, 4, 31, 58, 76, 84
maxillipeds, 84
May-beetles, 51, 52
May flies, 70, 72, 73
measly pork, 128
medulla, 300, 301
medullary sheath, 296
medusæ, 134, 135
meganucleus, 142, 143
Melanoplus, 18
Melanoplus femur-rubrum, 18
Melanoplus spretus, 18
Mendel, G., 373
Mendel's law, 373–378
Merino sheep, 356
mesenteries, 136
mesoderm, 336
mesothorax, 4
metabolism, 246
metacarpals, 290
metamorphosis, 337
 of amphibians, 175
 of echinoderms, 113
 of insects, 11, 12, 20, 31, 35, 51
metatarsals, 291
metathorax, 4
Meylan, Dr., 315
mice, 216
micronucleus, 142, 143
migration, of birds, 199–201
 of fishes, 162, 163
milk, souring of, 319
milkweed butterfly, 23
millipeds, 75, 76
mimmicry, 23–25, 364

INDEX 397

minnows, 167
mites, 81–83
molecules, 233, 234
moles, 216
molluscs, 98–107
molt, 21
molting, 10, 11, 21, 71, 72, 89
monarch butterfly, 23
monkeys, 227, 228
Monotremes, 211, 212
moose, 222
morphology, 232, 355
mosquitoes, 45–48, 148, 150
moths, 25–30
motor area, 302, 303
motor nerves, 296–298
motor reflex, 144
Musca domestica, 40, 41, 42
muscles, 243, 291–293
Muscular tissue, 243
mussels, 102
mutations, 356
Mya arenaria, 102
Myriapoda, 75, 76

N

Nam Family, 380
narcotics, 257
natural selection, 360–362
nauplius, 94
nautilus, 105, 106
Necturus, 170
nematocysts, 131, 136
nematodes, 122–124
nephridia, 117
nerve cells, 296
nerve cord, of Amphioxus, 155, 156
  of Balanoglossus, 155
  of crayfish, 88
  of earthworm, 118, 156
  of grasshopper, 8, 9
  of tunicate larva, 154
nerves, 294–299

nervous system, of Amphioxus, 155, 156
  of clam, 100, 101
  of crayfish, 88
  of earthworm, 118
  of grasshopper, 8, 9
  of man, 294–303
nervous tissue, 244
nests, 189, 191–194, 197, 338, 339
nettling cells, 131
newts, 169
nitrogen, 234, 236, 237, 238
Noctuidæ, 28
notochord, 154–159
Notonecta, 33
Novius (Vedalia) cardinalis, 38, 54
nuclei, 142, 241

O

ocelli, 3
Octopus, 107
oil glands, 283, 284
olfactory pits, 3
operculum, 158–160
Ophidia, 176
Ophiurans, 110
opossum, 214
orang-utans, 227
organs, 240
Ornithorynchus, 211, 212
orthoceratites, 105
Orthoptera, 1–16, 18, 19, 51
osculum, 139
ossicles, 87
ostia, 88
ova, 333
ovaries, 9
oviduct, 9
ovipositor, 5, 14, 19
owls, 204, 206, 207
ox, 220
oxidation, 235, 239, 270
oxygen, 234, 235–239, 260, 269, 271, 273, 275, 276

oyster drill, 105
oysters, 102, 105, 110

P

pain, 305
palp, 3, 76
pancreas, 250, 251, 281, 282
Pandorina, 331, 333
paper nautilus, 105
Paramœcium 141-144, 331, 332
parasites, 124-129, 144, 364, 366
parasitism, 96, 364-366
Parkes, Dr., 311
parthenogenesis, 35, 36, 333
Pasteur, L., 149, 319, 327
patella, 291
patent medicines, 330
Peabody, J. E., 316
pearls, 101, 102
pebrine, 149
Peckham, G. W., 64
Pecten, 102
pectoral arch, 160
pectoral fins, 158
pedal ganglia, 101
pedicellariæ, 110
pelvic arch, 160, 176
pelvic fins, 158
pelvis, 290
pepsin, 238, 249
peptone, 249
perch, 158, 167
pericardium, 88, 100
Perissodactyla, 220
phagocytes, 326
phalanges, 290, 291
pharynx, 116, 118, 154, 272
Phasmidæ, 19
phosphorus, 236, 238
phrenology, 302, 303
Phylloxera, 36
phylum, 18, 19
Physa, 104
physical changes, 232

physics, 233
physiology, 232
pig, 220
pigeon, 301, 357, 358
Pill bugs, 94
pineal gland, 346
Pinnipedia, 224
Pithecanthropus erectus, 355
placenta, 214, 335
plague, 49, 50
planarians, 124
Planorbis, 104
plant lice, 35, 36
plasma, 259
Plasmodium, 148
pleura, 273
pleurisy, 274
pluteus, 113
pneumonia, 314
poison gland, 76
pollen basket, 59
pollen combs, 59
polyp, 136-138
pond skaters, 34
porcupines, 209, 216
Porifera, 139
portal vein, 266
Portuguese man-of-war, 135, 136
potassium, 234, 236, 238
prawns, 90
Primates, 227-230
Proboscidea, 220
proglottids, 126, 127
pro-legs, 21
propolis, 62
prostomium, 114
protective resemblance, 25, 363, 364
proteins, 237, 238, 249, 251, 252, 253, 255
Proteus, 170
prothorax, 4
protoplasm, 238, 252
Protozoa, 141-152, 317, 331

Psammophila, 64
pseudopodia, 140, 146
ptarmigan, 184
pterodactyls, 183
ptyalin, 238, 248
pulmonary arteries, 263
pulmonary veins, 263
puma, 223, 224
pupæ, of insects, 12, 21, 22, 26, 41-43, 45, 46, 49, 51, 52, 62, 66, 67
pupil, 307
pus, 261
pylorus, 249
pythons, 176, 179

Q

queen ant, 66, 67
queen bee, 56-58, 60-63

R

rabbits, 360
rabies, 326, 327
Radiates, 108
Radiolaria, 148, 149
radius, 290
Rana catesbiana, 172, 173
Rana pipiens, 172
Ranidæ, 172
rats, 49, 50, 216
rattlesnakes, 177, 178
Ray Lankester, E., 76
rays, 158, 164, 165
rectum, 7, 249
red-legged grasshopper, 18
reflex acts, 297-299
regeneration, of amphibians, 170
    of earthworm, 119
    of Hydra, 133
    of starfish, 110
reindeer, 222
renal artery, 277, 278
renal vein, 277, 278

reproduction, 331-340
    in Amœba, 147, 331
    in amphibians, 171
    in ants, 67
    in aphids, 35, 36
    in bees, 61-63
    in birds, 188, 194, 338, 339
    in clam, 101
    in crayfish, 89
    in earthworm, 118, 119
    in fishes, 160-164, 338
    in flies and mosquitoes, 41-46
    in grasshopper, 9, 10
    in hookworm, 123
    in Hydra, 132
    in jelly fish, 134, 135
    in liver fluke, 125, 126
    in mammals, 335
    in Paramœcium, 143, 331, 332
    in spiders, 80
    in sponges, 140
    in tapeworm, 127-129
    in Trichina, 123
    in Volvox, 332, 333
reptiles, 176-182, 338, 349, 350, 352
respiration, 239
    in amphibians, 170, 172
    in birds, 188
    in clam, 100
    in crayfish, 86
    in fishes, 159, 160
    in grasshopper, 5
    in man, 269-276
    in Protozoa, 151
    in snails, 103, 104
    in spiders, 77
retina, 307, 308
reversion, 370
ribs, 274, 288, 290
Roberts, 312
robin, 119, 191-194, 207
Rocky Mountain fever, 82

INDEX

Rocky Mountain locust, 12, 13, 18
Rodentia, 216, 217
rostrum, 84
round worms, 122–124
rudimentary organs, 343–346
ruminants, 220

S

Sacculina, 96, 97, 365
sacrum, 290
salamanders, 169, 170, 364
salivary glands, 248
salmon, 162, 163
salts, 256
San José scale, 37
sand dollars, 112
saprophytes, 144
Sarcodina, 141, 146–148
scale bugs, 36–38
scales, 159, 176, 177, 183
scallop, 102
scapula, 290
sclerotic, 306
scorpions, 76, 81
screw worm, 42
sea anemones, 130, 136
sea cows, 226, 227
sea cucumbers, 112
sea-horse, 167
sea-lions, 224, 225
sea urchins, 110–112
seals, 224, 225
Sebaceous glands, 283, 284
segments, 126
semicircular canals, 310
sensations, 294, 297, 304–310
sense organs, 304–310
sensory nerves, 296–298
septa, 116
serum, 260, 283
setæ, 115
seventeen-year cicada, 35
sexual selection, 195, 196
sharks, 164, 165

sheep, 220, 223
shell, 98
shrews, 216
shrimps, 90, 91
silk-worm moth, 26, 27, 149
Silurian period, 349
sinuses, 88
siphon, 99, 102, 106
Siphonaptera, 49
Siphonophores, 135
Sirenia, 226
skates, 164, 165
skeleton, 188, 287–291
skin, 277, 279, 283–286
skull, 160, 289
skunk, 210
sleep, 303
sleeping sickness, 146
smallpox, 325, 328
smell, 3, 209, 210, 305
snails, 98, 103, 104
snakes, 176
sneezing, 299
sodium, 234, 236
somites, 84
sow bugs, 94
species, 17, 18, 341, 342
Spencer, H., 313
sperm cells, 160, 333, 334
spermaceti, 226
spermaries, 9, 10
spermatozoa, 333, 334
   of earthworm, 118, 140
   of fishes, 160
   of grasshopper, 10
sphinx moths, 27, 28
spiders, 76–81
spiracles, 5
sponges, 130, 138, 139, 140
Spongilla, 140
spores, 141, 147, 148, 331
Sporozoa, 141, 148–150
sports, 356
squash bug, 31

INDEX 401

squid, 106, 107
squirrels, 216
starfishes, 108–110
Stegomyia, 48
sternum, 188, 274, 290
stimulants, 256
sting. 60
sting-rays, 165
stirrup, 310
Stockard, C. R., 383
stomach, of grasshopper, 6
    of clam, 100, 101
    of crayfish, 87
    of man, 248, 249, 254
    of starfish, 109, 110
stone flies, 70, 73
sturgeon, 166
subesophageal ganglia, in crayfish, 88
    in grasshopper, 8, 9
    of clam, 99–101
Sullivan, Dr., 313
sulphur, 234, 236, 238
sunfish, 158
suprarenal bodies, 281
swimming bladder, 160
symbiosis, 133, 366, 367
sympathetic system, 296
synovial membrane, 287
systemic circulation, 263

T

Tabanidæ, 44
Tachinidæ, 44
Tænia solium, 127, 128
tail, 184, 185, 195, 196
tail fin, 158
tapeworm, 126–129
tarsals, 291
tarsus, 4
taste buds, 305
tea, 257
teeth, 177, 247, 248, 346

Teleostomi, 164, 166
teleosts, 167
telson, 84, 85
tendons, 243, 291, 292
tentacles, 103, 130, 132, 134, 135
testes, 9, 10
Texas fever, 82, 148, 317
thoracic duct, 266
thorax, of crab, 91
    of crayfish, 84–88
    of grasshopper, 2, 4
thyroid extract, 281
thyroid gland, 281
tibia, of grasshopper, 4
    of man, 290
ticks, 81, 82, 149
tigers, 223
tissues, 240, 241–244
toads, 169, 171, 173–175
tobacco, 257, 268, 314–316
tonsils, 267
torpedo, 166
tortoise, 176, 181, 182
touch, 3, 304
toxin, 326
trachea, 272
tracheæ, 5, 77
tracheal gills, 72
tree frogs, 174, 175
Trematodes, 124–126
Trichina, 122, 123
Trichinella spiralis, 122, 123
trichinosis, 317
Triton, 170
trypanosomes, 146
trypsin, 251
tsetse-fly, 146
tube feet, 108, 109, 112
tuberculosis, 314, 322–324, 377
tunicates, 153–155, 334
Turbellaria, 124
turtles, 176, 181, 182
tympanic membrane, 309
typhoid fever, 324–326, 328, 329

U

ulna, 290
Ungulata, 220
urea, 278
ureter, 277, 278
urinary tubules, 7, 8, 277, 278
Urodeles, 169
Ursus americanus, 18
Ursus arctos, 18
Ursus horribilis, 18, 224

V

vaccination, 328
vagus nerve, 300
valves of heart, 7, 88, 262
  of shell, 98
  of veins, 265
variation, 356, 372
varieties, 17, 356-359
veins, 5, 262-266
ventricles, 100, 262
vermiform appendix, 250, 344, 345
vertebræ, 160, 289, 290
vertebral column, 160, 289
Vertebrates, 153-157
viceroy butterfly, 23
villi, 250
visceral ganglia, 100
vision, 3, 307, 308
vitreous humor, 307
vocal sacs, 171
Volvox, 332, 333
vultures, 202

W

walking-stick, 16
Wallace, A. R., 358
walruses, 224
warble flies, 43
warning colors, 170, 364
wasps, 56, 64, 65
water fleas, 97
water moccasin, 179
water scorpion, 33
water striders, 34
wax glands, 60
wax pincers, 59
web, 77-80
weevils, 55
Weismann, A., 370
whalebone, 226
whales, 225, 226, 346
Whetham, 385
whirling beetles, 54
White, A. D., 315
white corpuscles, see *Leucocytes*
Wiedersheim, R., 346
windpipe, 272
winking, 299
wolves, 223
worker bees, 56-62
worms, 114-121
wrigglers, 45-48

Y

yellow fever, 48
yellow jackets, 65

CPSIA information can be obtained
at www.ICGtesting.com
Printed in the USA
BVHW07*2040200818
525056BV00013B/1449/P